T0135520

Oil Film Dynamics in Aero Engine Bearing Chambers - Fundamental Investigations and Numerical Modelling

Zur Erlangung des akademischen Grades eines

Doktors der Ingenieurwissenschaften

der Fakultät für Maschinenbau
Karlsruher Institut für Technologie (KIT)

genehmigte

Dissertation

von

MSc.-Ing. Amir Aleem Hashmi
aus Karachi, Pakistan

<table>
<tr><td>Tag der mündlichen Prüfung:</td><td>19.07.2012</td></tr>
<tr><td>Hauptreferent:</td><td>Prof. Dr.-Ing. H.-J. Bauer, Ord.</td></tr>
<tr><td>Korreferent:</td><td>Prof. Dr.-Ing. T. Schulenberg</td></tr>
</table>

Forschungsberichte aus dem
Institut für Thermische Strömungsmaschinen

herausgegeben von:
Prof. Dr.-Ing. Hans-Jörg Bauer,
Prof. Dr.-Ing. Dr.-Ing. E.h. Dr. h.c. mult. Sigmar Wittig
Lehrstuhl und Institut für Thermische Strömungsmaschinen
Karlsruher Institut für Technologie (KIT)
Kaiserstr. 12
D-76128 Karlsruhe

Bibliografische Information der Deutschen Nationalbibliothek

Die Deutsche Nationalbibliothek verzeichnet diese Publikation in der
Deutschen Nationalbibliografie; detaillierte bibliografische Daten sind
im Internet über http://dnb.d-nb.de abrufbar.

ISSN 1615-4983
ISBN 978-3-8325-3228-4

Logos Verlag Berlin GmbH
Comeniushof, Gubener Str. 47,
10243 Berlin
Tel.: +49 030 42 85 10 90
Fax: +49 030 42 85 10 92
INTERNET: http://www.logos-verlag.de

Oil Film Dynamics in Aero Engine Bearing Chambers - Fundamental Investigations and Numerical Modelling

von

MSc.-Ing. Amir Aleem Hashmi

Karlsruhe 2012

Vorwort des Herausgebers

Der schnelle technische Fortschritt im Turbomaschinenbau, der durch extreme technische Forderungen und starken internationalen Wettbewerb geprägt ist, verlangt einen effizienten Austausch und die Diskussion von Fachwissen und Erfahrung zwischen Universitäten und industriellen Partnern. Mit der vorliegenden Reihe haben wir versucht, ein Forum zu schaffen, das neben unseren Publikationen in Fachzeitschriften die aktuellen Forschungsergebnisse des Instituts für Thermische Strömungsmaschinen am Karlsruher Institut für Technologie (KIT) einem möglichst großen Kreis von Fachkollegen aus der Wissenschaft und vor allem auch der Praxis zugänglich macht und den Wissenstransfer intensiviert und beschleunigt.

Flugtriebwerke, stationäre Gasturbinen, Turbolader und Verdichter sind im Verbund mit den zugehörigen Anlagen faszinierende Anwendungsbereiche. Es ist nur natürlich, dass die methodischen Lösungsansätze, die neuen Messtechniken, die Laboranlagen auch zur Lösung von Problemstellungen in anderen Gebieten - hier denke ich an Otto- und Dieselmotoren, elektrische Antriebe und zahlreiche weitere Anwendungen - genutzt werden. Die effiziente, umweltfreundliche und zuverlässige Umsetzung von Energie führt zu Fragen der ein- und mehrphasigen Strömung, der Verbrennung und der Schadstoffbildung, des Wärmeübergangs sowie des Verhaltens metallischer und keramischer Materialien und Verbundwerkstoffe. Sie stehen im Mittelpunkt ausgedehnter theoretischer und experimenteller Arbeiten, die im Rahmen nationaler und internationaler Forschungsprogramme in Kooperation mit Partnern aus Industrie, Universitäten und anderen Forschungseinrichtungen durchgeführt werden.

Es sollte nicht unerwähnt bleiben, dass alle Arbeiten durch enge Kooperation innerhalb des Instituts geprägt sind. Nicht ohne Grund ist der Beitrag der Werkstätten, der Technik-, der Rechner- und Verwaltungsabteilungen besonders hervorzuheben. Diplomanden und Hilfsassistenten tragen mit ihren Ideen Wesentliches bei, und natürlich ist es der stets freundschaftlich fordernde wissenschaftliche Austausch zwischen den Forschergruppen des Instituts, der zur gleichbleibend hohen Qualität der Arbeiten entscheidend beiträgt. Dabei sind wir für die Unterstützung unserer Förderer außerordentlich dankbar.

Im vorliegenden Band der Veröffentlichungsreihe befasst sich Herr Hashmi in Fortführung früherer Arbeiten des Instituts mit der numerischen Beschreibung schubspannungsgetriebener Flüssigkeitsfilme. Konkreter Anwendungsfall sind Ölfilme in Lagerkammern von Flugtriebwerken, die neben der durch die Rotation der Welle induzierten Luftströmung Schwerkrafteinflüssen ausgesetzt sind. Für die numerische Beschreibung dieser Filme, die elliptischen Charakter besitzen, können Modelle, wie sie in der Vergangenheit für die Beschreibung parabolischer dünner Filme entwickelt wurden, nicht verwendet werden. Die vorliegende Arbeit verfolgt neuartige Ansätze zur Beschreibung dicker schubspannungs- und schwerkraftgetriebener Flüssigkeitsfilme und ihre experimentelle Verifikation. Bei der zur numerischen Beschreibung schubspannungsgetriebener Flüssigkeitsfilme grundsätzlich geeigneten Volume Of Fluid (VOF) Methode deckt Herr Hashmi ein grundsätzliches Problem existierender Verfahren bei der Behandlung der Grenzfläche für die turbulenten Feldgrößen auf. Er entwickelt einen neuen Ansatz, der den turbulenten Impulsfluss korrekt beschreibt. Zur experimentellen Verifikation setzt Herr Hashmi eine generische Filmversuchsstrecke ein, die die Untersuchung sowohl von Schwerkraft- als auch Schubspannungseinflüssen auf die Filmströmung ermöglicht. Als einfaches aber aussagekräftiges integrales Maß betrachtet er den durch die wellige Grenzfläche verursachten zusätzlichen Druckverlust der Luftströmung. Die Ergebnisse zeigen klar die Überlegenheit des neu entwickelten Ansatzes gegenüber der Standard-VOF-Methode.

Karlsruhe, im September 2012   Hans-Jörg Bauer

Vorwort des Autors

Die vorliegende Arbeit entstand während meiner Tätigkeit am Institut für Thermische Strömungsmaschinen.

Ich möchte dem Leiter des Instituts Herrn Prof. Dr.-Ing. Hans-Jörg Bauer meinen besonderen Dank für seine wertvolle Betreuung im Verlauf der Arbeit aussprechen. Nur seine beständige Motivation und die finanzielle Unterstützung durch andere Projekte ermöglichte es mir die Dissertation pünktlich fertigzustellen.

Ich möchte auch Herrn Prof. Dr.-Ing. Thomas Schulenberg danken, der die Rolle des Korreferats der Arbeit übernommen hat. Insbesondere danke ich ihm für die mehrmaligen Treffen in denen er viele Ideen mit mir teilte.

Der Rolle von Herrn Dr.-Ing. Klaus Dullenkopf als direkten Betreuer, der mir bei Unwegbarkeiten stets zur Seite stand, gilt meine besondere Wertschätzung. Der Hilfestellung von Herrn Dr.-Ing. Rainer Koch bei der Findung der richtigen Richtung und für die erfolgreiche Fertigstellung der Arbeit lässt sich nur schwer in Worten danken. Die gelegentlichen Diskussionen und der Gedankenaustausch mit anderen Kollegen des Instituts, insbesondere Herrn Dipl.-Ing. Davide Peduto und Herrn Dipl.-Ing. Martin Schwitzke, stärkten meine Motivation und halfen mir die optimalen Lösungen zu finden.

Der experimentelle Teil meiner Arbeit wäre nicht möglich gewesen ohne die wertvolle Unterstützung der Werkstätten, besonders von den Leitern Herrn Günther Jettke und Herrn Wolfgang Schimmeyer. Sie waren stets bereit Lösungsansätze für die technischen Problemstellungen zu finden und hielten alles am Laufen.

Die Bedeutung von Frau Viola für das Institut ist schwer zu greifen. Durch ihre stete Zuarbeit und ihre Hilfestellung bei der Bewältigung kleiner Unwegbarkeiten für jeden am Institut, war sie auch mir bei vielen Gelegenheiten eine große Unterstützung. So möchte ich an dieser Stelle Ihre wertvollen Tipps auf der Suche nach verloren gegangenen Daten erwähnen, sowie ihre Organisation der Feierlichkeiten im Rahmen der Doktorprüfung und am Sommerfest.

Da sich ein großer Teil der vorliegenden Arbeit mit der numerischen Modellierung beschäftigt, war ein betriebsbereites Rechnersystem und die Gewährleistung zuverlässiger Datensicherung ein großer Gewinn. Dafür möchte ich Herrn Michael Lahm danken, der meinen Rechner am Laufen hielt und Daten sicherte, sowie für die schnelle Hilfestellung bei Hard- und Software Problemen, Erneuerungen und Instandhaltung.

Ohne ausreichende Unterstützung und Verständnis von Seiten der Familie und Freunde ist es schwer im Leben erfolgreich zu sein. Deshalb möchte ich mich für deren Unterstützung und Verständnis bedanken in den Zeiten in denen ich diese benötigte. Dies gilt besonders für meine Frau, die mich stets mit Ihrer Liebe und Fürsorge unterstützt hat ohne nachzufragen, weshalb ich so spät vom Büro nach Hause gekommen bin.

Karlsruhe, im September 2012   Amir Aleem Hashmi

Contents

Nomenclature

Symbol	Unit	Quantity
A	$-, m^2, 1/m$	Variable/Area/ Interfacial Area Density
C	$-, J/kg.k$	Constant/Specific heat
$C_{1\varepsilon}, C_{2\varepsilon}, C_\mu$	$-$	Turbulence model constants
D	$-$	Turbulence model destruction terms
E	$1/s$	Strain tensor
f	$-, Hz$	Friction factor/ Frequency
F	N	Force
g	m/s^2	Acceleration due to gravity
h	m	Film thickness
H	m	Height
I	Pa	Additional momentum transfer to film
k	$m^2/s^2, W/m.k$	Turbulent kinetic energy/Thermal conductivity
K	m^2/s^2	Mean flow kinetic energy
l	m	Characteristic length
$\dot{m}$	$kg/m^3 s$	Mass source/sink due to phase change
M	N	Momentum source
n	$-, rev/min$	Normal vector/Revolution per minute
P	$Pa, kg/ms^3$	Pressure/Production terms in a 2-Eq. turbulence model
S	$kg/m^3 s$	Explicit mass source/sink
t	sec	Time
u, v, w	m/s	Velocity components in space
u^+	$-$	Dimensionless velocity
U	m/s	(Mean) Velocity vector
$\dot{V}$	$m^3/s, l/min$	Film loading
x, y, z	m	Space coordinates
y^+	$-$	Dimensionless distance to wall

Greek Symbols

α	$-$	Volume fraction/Turbulence model constant
β	$-$	Turbulence model constant
β^*	$-$	Turbulence model constant
γ	$-$	Phase indicator function

λ	$-$	Switching function
φ	$(^\circ)$	Angle
ϕ	$-$	Any flow field variable
θ	$(^\circ)$	Angle/Wall contact angle
ε	$m^2/s^3,-$	Eddy dissipation
ω	$1/s$	Eddy frequency
ℓ	m	Length scale of the turbulent flow field
μ	kg/ms	Dynamic viscosity
ν	m^2/s	Kinematic viscosity
ρ	kg/m^3	Density
σ	N/m	Surface tension
δ	$m,-$	Film thickness/Interface Kronecker delta
ϑ	m/s	Velocity scale of the turbulent flow field
$\sigma_{k,\varepsilon,\omega}$	$-$	Prandtl number
Δ	$-$	Delta (difference)
τ	N/m^2	Stress
$\Gamma_t = \frac{\mu_t}{\sigma_{k,\varepsilon,\omega}}$	kg/ms	Turbulent diffusivity
Γ	kg/m^3s	Mass source/sink (explicit or due to phase change)

Indices

$^{-}$	Average
$^{\wedge}$	Effective
$'$	Fluctuating component
$*$	Non dimensionalized quantity
f,F	Film
G	Gas
i	Interface
i,j	Indices of the Cartesian tensor notation
L	Liquid
q	Phase indicator
S	Superficial
St	Static
t	Turbulent
T	Total
w	Wall
WF	With film
WoF	Without film

Dimensionless Numbers

$C = \frac{u \Delta t}{\Delta x}$ Courant number

$Fo = \frac{kt}{\rho C_p l^2}$ Fourier number

$Fr = \frac{U^2}{\sqrt{gl}}$ Froude's number

$Re = \frac{Ul}{\nu}$ Reynold's number

$We = \frac{\rho U^2 l}{\sigma}$ Weber number

Abbreviations

BC	Bearing Chamber
CCFL	Counter-Current Flow Limitation
CFD	Computational Fluid Dynamics
CSF	Continuum Surface Force
EXP	Experiment
FFT	Fast Fourier Transform
HP	High Pressure
HSBC	High Speed Bearing Chamber
HWA	Hot Wire Anemometry
IP	Intermediate Pressure
ITS	Institut für Thermische Strömungsmaschinen
LDA	Laser Doppler Anemometry
LDV	Laser Doppler Velocimetry
LOCA	Loss-Of-Coolant Accident
LP	Low Pressure
MAC	Marker And Cell
PDA	Photochromatic Dye Activation
PDF	Probability Density Function
PSD	Power Spectral Density
SR	Scavenge Ratio
VOF	Volume-Of-Fluid

1 Introduction - Bearing Chambers in Aero Engines

The load bearing capacity and almost a failsafe running function of the rolling element bearings makes them an integral part of modern aero engines. Based on these characteristics, the rolling element bearings cannot be replaced in aero engines within the near future. However, due to bearing friction at aero engine specific shaft speeds, the rolling element bearings have excessive heat generation and special efforts have to be directed to this problem. The oil system in aero engines is designed to carry out the tasks of lubrication and cooling, where the oil volume necessary for cooling is considerably larger than it is required for lubrication (Schmidt et al. (1982)). Depending on the engine type, the cooling/lubricating oil is either injected beneath (under race lubrication) or in front of the bearings. In order to prevent oil fires, the oil must be kept separated from the hot surroundings. This is achieved by building chamber(s) around the bearing(s) hence the name bearing chamber. Near the rotating shaft, the bearing chambers are generally sealed with the help of contactless labyrinth seals pressurized by air taken from the compressor bleeds (see Fig. 1.1). The fast moving shaft accelerates the air in the circumferential direction and in the process influences the oil flow inside the bearing chambers. As a result of interaction between the cooling oil flow and sealing air, an extremely complex two-phase flow consisting of droplets laden air core and an oil film attached to the outer bearing chamber walls dominates the bearing chambers (Glahn (1995)). To keep the bearing chambers reliably functioning, the oil has to be continuously replaced by the new cooler oil. Due to high pressure in the chambers, the hot air/oil mixture leaves the bearing chamber system through the connected vent and scavenges pipes. In some engines, an oil pump is provided to the scavenge pipe to assist the recirculation of the oil. Before the oil can be pumped again to the bearings, the air/oil mixture runs through a breather where after separation oil is directed to the oil tank and air is released to the environment (see Fig. 1.2).

Amongst the design criteria for modern aero engines (i.e. high turbine inlet temperatures and shaft speeds), safety and reliability aspects are becoming more and more important. To ensure a high level of safety and reliability for several thousands of operating hours, the oil system must be designed perfectly. In the present, experience of the manufacturers and simplified correlations play a major role in the design process of bearing chambers and their off-take geometries. This however, limits the capability to cope with the higher process temperatures and counter-rotating rotor concepts. For the bearing chamber located near the combustor and turbine, the contact of the piping with the hot gases (with temperatures well above the oil coking temperature) cannot be avoided. This scenario can become even more alarming, due to the accumulated heat, if the engine shuts down immediately. Therefore, to avoid the risk of oil coking or oil fire, the temperature of the oil must be kept as low as possible. This proofs the necessity of exact temperature information for any attempt to increase the efficiency without compromising the reliability of the system. This in turns points out the need to precisely understand the complex two-phase flow phenomena in the aero engine bearing chambers.

At the Institut für Thermische Strömungsmaschinen, bearing chambers is a topic of active research since the early 1990's. Most comprehensive is the work of Glahn (1995) who characterized the two-phase flow phenomena using a model bearing chamber test rig. Existence of droplet rich gas core was proved. The droplet size and velocity spectrum at one location was measured. An oil film attached to the chamber wall was detected for all investigated test conditions. In all later investigations and publications of this area by various authors e.g. Chandra et al. (2010) and particularly in experimental observations made at the ITS (Wittig et al. (1994), Glahn (1995), Glahn und Wittig (1996), Schmälzle (1997), Gorse et al. (2007)), an oil film attached to the outer wall of the bearing chamber was confirmed. The oil film is also necessary to keep the outer wall temperature within permissible limits; however, oil fire or coking may result if the residence time of oil present in the bearing chambers exceeds a certain limit. Therefore, any system optimization with regard to the most important design parameters of a bearing chamber, efficient lubrication, reliable heat management and efficient scavenging, requires a detailed knowledge on the oil film dynamics.

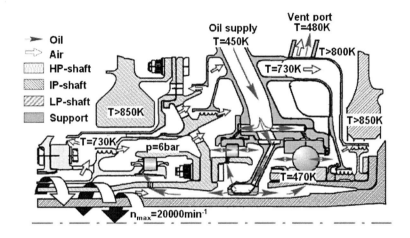

Fig. 1.1: Rear bearing chamber of a three spool aero engine (Glahn (1995))

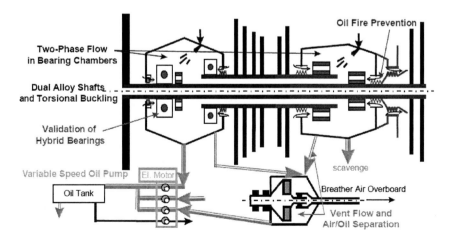

Fig. 1.2: Schematics of a typical oil system (Klingsporn (2004))

In the last two decades, a large number of experimental works have been performed using simple (Chandra et al. (2010)) and model bearing chamber test rigs (Wittig et al. (1994), Glahn (1995), Schmälzle (1997), Gorse (2007)) to understand the complex two-phase flow phenomena occurring in a typical bearing chamber environment. However, due to the complexity of the multiphase flow inside the bearing chambers no universally valid design rules could be delivered. The difficulty lies in the existence of numerous possible flow patterns and phase interactions which cannot be, or only with highly expensive experimental setups, determined in real engine conditions. The limitations on instrumentation and optical access in bearing chambers together with the fact that coexisting flow phenomena (dropletfilm interaction, off-take disturbances, continuous film loading from all possible circumferential positions etc...) cannot be avoided, makes it very difficult to sufficiently resolve the individual characteristics of the flow field (wall film, droplets, off-take flow). Especially in the vicinity of the scavenge off-takes, where the oil film is relatively thick and highly dynamic in nature, no comprehensive understandings could be delivered till today. In addition to that, any quantitative analysis of the film is usually very sensitive to the choice of boundary conditions which in 1:1 bearing chamber test rigs are very difficult to determine accurately. The growing interest in CFD technique also requires well-defined boundary conditions and understanding of individual film characteristics before a superposition of all on the film dynamics (e.g. in bearing chambers) can be simulated. On the other hand, as will be elaborated in the next chapter, laboratory data using simple test rigs are scarce and very often not fully relevant as well as being difficult to apply. Therefore, test rigs reproducing bearing chamber typical wall film flows with well-defined boundary conditions are necessary not only for analytical/empirical treatments but also to analyse the available multiphase models for oil film simulations. A classical similarity is however difficult to achieve due to the fact that a non-dimensionless analysis requires a prior knowledge on several unknowns. Consequently, a simple test rig ensuring similar conditions to the ITS model bearing chamber test rig is designed and serves the purpose of investigating the fundamentals of isolated bearing chamber films with focus on thick film dynamics.

For any change in a bearing chamber configuration, the flow field usually changes significantly. Therefore, it is very difficult to ascertain that the data obtained for a certain model bearing chamber test rig is equally valid for another bearing chamber configuration. In this case, CFD can be a potential tool for the design and optimization of bearing chambers. The investigation of flow field can be dealt with less effort than required for experimental investigations e.g. individual characteristics of the flow field can be studied efficiently and effectively. A parameter study in this regard will reveal the major features of the flow field and sensitivity to geometric changes can be studied. Regarding the film flow in bearing chambers, in the past, a rimming flow was assumed. Hence, it was assumed that the interfacial shear stress is the major driving force on the film and gravity has only a limited influence on the film dynamics even in the counter-current regime. Consequently, several thin film models were proposed in the past e.g. Glahn et al. (1996), Chew (1996), Farrall et al. (2004), Gorse et al. (2004) and Shimo et al. (2005) but either no or only limited success could be achieved especially where the film is comparatively thicker. The fact that oil film is continuously being fed from all possible circumferential positions and also being continuously removed from the off-takes, a clean rimming flow at all times and shaft speeds is very unlikely. Furthermore, the aero engine manufacturers still experience the problem of oil coking which is a direct result of increase in residence time in the bearing chambers. The recirculation zone on the counter-current side in the vicinity of the scavenge port is one of the modes of increase in residence time. Schmälzle (1997), Gorse (2007), Peduto & Kurz (2010) indeed observed a partial flow reversal (recirculation zone) with local film thickening in the counter-current regime for some geometries and chamber dimensions. This behaviour of film cannot be reproduced by the thin film models due to a fundamental model limitation (equation set, instead of elliptic, is parabolic in nature). Therefore, enhanced and more detailed "thick" film models are necessary. For this purpose, multiphase modelling techniques available in modern CFD codes need to be analysed for their applicability to the bearing chamber typical thick film flows. The free surface model, commonly known as the Volume-Of-Fluid method, can be a potential tool which can also deal with heat transfer problem. Plausible information on the heat transfer will help design better (smaller and lighter) coolers, which means higher efficiency and reduced weight. Known oil/air ratio and temperature distribution in the off-takes will also help making measures against coking more effective. Avoiding coking and oil fire effectively means gains in the safety and reliability of the system. This information will also help in developing better correlations for the heat transfer thus reducing the development time of the engines (Zimmermann et al. (1991)). Hence, cost reductions will occur during the development phase and during the operation (increased payload by weight reduction). Therefore, in this thesis, the potential offered by the state of the art multiphase CFD technique is explored for its applicability.

2 Oil Film Dynamics - State of the Art

As mentioned in the last chapter, due to the importance of bearing chambers in the oil system, several experimental studies using simple and model bearing chamber test rigs have been conducted to understand the relevant two-phase flow phenomena. For almost the last two decades, the Institut für Thermische Strömungsmaschinen is also providing state of the art expertise in the development of aero engine bearing chambers. Besides the investigation of complete bearing chamber flow fields (Glahn (1995)), individual bearing chamber features like off-take geometries (Schober (1998)), off-take flow phenomena (Busam (2004)), droplet generation (Gorse (2007)) etc... have also become a topic of large number of research works conducted at the institute. The analysis of film flow in bearing chambers has also received much of the needed attention dating back to the early 1990's when Wittig et al. (1994) determined for the first time the film thickness and heat transfer under engine relevant conditions. Many other comprehensive experimental studies (e.g. Glahn (1995, 1996), Schmälzle (1997), Lipp (2003) etc...) have also followed to investigate the oil film dynamics in bearing chambers. The quantitative analysis was mostly based on the film thickness and velocity profile measurements. Owing to the importance of oil film dynamics, several numerical attempts have also been made to understand the underlying physical phenomena. In the following, the relevant experimental and numerical investigations conducted in sophisticated as well as simplified test rigs reproducing bearing chamber typical film phenomena are reviewed and the needs to complement these investigations are identified.

2.1 Experimental Investigations

2.1.1 ITS High Speed Bearing Chamber (HSBC) Test Rig

In Fig. 2.1, the cross-sectional view of the Institut für Thermische Strömungsmaschinen (ITS) high speed rotating bearing chamber test rig is shown. The test rig represents a model bearing chamber where real engine conditions for high as well as intermediate pressure bearing chambers can be simulated (Wittig et al. (1994)). Sealing air and lubrication oil flow could be varied nearly in the whole range of aero engine applications. Due to the compact and modular design, variations of bearing chamber geometries and roller bearings can be realized with relatively low financial and manufacturing effort. Detailed design considerations and a description of all test facility components are given in Wittig et al. (1994). However, for a better understanding of air and oil flows inside the test rig, a short description of the general arrangement is reproduced from Gorse et al. (2004).

The rotor is supported by two bearings which can be rotated up to $20,000rpm$. A ball bearing is used to prevent axial and radial displacements of the rotor. Due to the small size of the ball bearing only a little amount of oil is required for lubrication. The second bearing is a squeeze film damped roller bearing. Oil is supplied by an under-race lubrication system which has a maximum capacity of $400l/hr$. To simulate engine representative conditions, the inlet oil can be heated up to $100°C$. The roller bearing separates the two bearing chambers (see Fig. 2.1), where the oil is ejected after cooling and lubricating the bearing. Both chambers are sealed by air pressurized labyrinth seals. The sealing air is provided by a compressor with a maximum capacity of $0.5kg/s$ at a pressure ratio of $\Pi = 10$. The resulting air/oil mixture is discharged at the top and bottom of each chamber by the vent and scavenge system (not shown in Fig. 2.1), respectively. The oil is then separated from the air using a standard centrifugal filter unit and returned back to the tank.

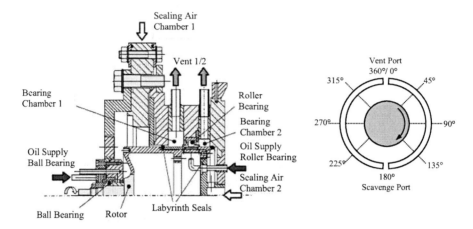

Fig. 2.1: ITS High Speed (model) Bearing Chamber test rig (Left); Probe locations (Right)

In the last two decades, the test rig was intensively used for a large number of studies in order to comprehend the two-phase flow phenomena in real aero engine bearing chambers. In the following, only the investigations dealing with the film flow dynamics are considered.

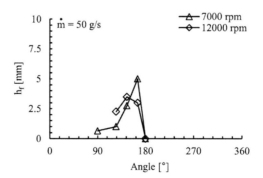

Fig. 2.2: Bearing chamber typical film thickness distributions (Wittig et al. (1994))

Wittig et al. (1994) succeeded in resolving the film thickness up to $15°$ upstream of the scavenge off-take and found that certain bearing chamber configurations can manifest film thicknesses as high as $5mm$ even at intermediate shaft speeds. The influence of shaft rotational speeds on the oil film thickness and local heat transfer was also investigated. The results revealed that gravitational

forces as well as the rotational speed of the shaft have a strong impact on the film height.
Glahn (1995) investigated the dependence of circumferential film thickness distributions on
the chamber geometry, shaft speed and oil feed. For this purpose, he employed two bearing
chamber configurations with different dimensions ($b \times h = 20 \times 28mm$ & $b \times h = 46 \times 28mm$)
and measured the film thickness distribution at five angular positions. The shaft speed was
varied in the range $4,000rpm \leq n \leq 16,000rpm$ and the oil feed was varied in the range
$50l/hr \leq \dot{V} \leq 200l/hr$. The influence of shaft speed on the oil film thickness was determined
by keeping a constant oil feed of $100l/hr$ and increasing the shaft speed in the given range. It
was observed that at least a shaft speed of $n \geq 12,000rpm$ was necessary to overcome the effect
of gravitation and to achieve a homogenously driven oil film as shown in Fig. 2.3 (left). The
inhomogeneity of film distribution in the first chamber geometry (Fig. 2.3 (left)) was stronger
than in the second geometry (Fig. 2.3 (right)). A difference in core gas velocities as suggested by
Gorse et al. (2004) can be stated as a reason for this behaviour.

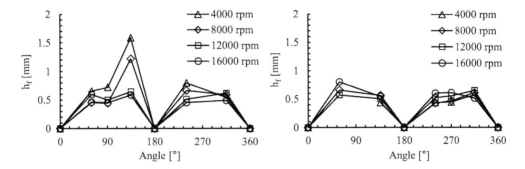

Fig. 2.3: Influence of shaft speed on the oil film distribution (Glahn (1995))
($b \times h = 20 \times 28mm$ (Left); $b \times h = 46 \times 28mm$ (Right))

By keeping a constant shaft speed of $12,000rpm$ and increasing the oil feed in the given range,
the influence of oil feed on the circumferential oil film thickness distribution was determined.

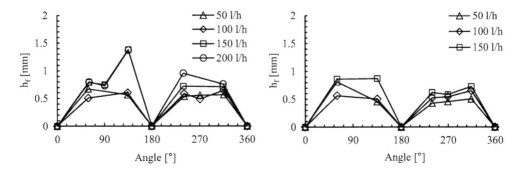

Fig. 2.4: Influence of oil feed on the oil film distribution (Glahn (1995))
($b \times h = 20 \times 28mm$ (Left); $b \times h = 46 \times 28mm$ (Right))

Analogous to the variable shaft speed investigations (Fig. 2.3), an inhomogeneous film distribution was achieved for the first chamber geometry (see Fig. 2.4 (left)) with the maximum film thickness of approx. $1.5mm$. The film thickness in the second chamber geometry remains more uniformly distributed (see Fig. 2.4 (right)) as was also the case in the variable speed investigations confirming a strong influence of the geometry. These investigations revealed that the chamber geometry can affect the film distribution independent of the operating conditions in a manner explained by Gorse et al. (2004) i.e. different core gas velocities may result for different chamber configurations. Glahn et al. (1996) used a fibre-optic LDV setup for the measurements of film velocity profiles at a point located 30° upstream of the scavenge port (in the co-current side). The test rig developed by Wittig et al. (1994) was adapted to the velocity profile measurements; hence,real engine conditions could be simulated. Oil and sealing air flows were kept constant at $150l/min$ and $10g/s$ respectively. The shaft rotational speed was varied in the range $3,000-12,000rpm$ and its effect on the film thickness variation was studied. It was found that increasing the rotational speed resulted in a decreased film thickness; consequently, an increase in the film velocity (Fig. 2.5). This was a direct result of an increase in the interfacial shear stress on the gas-liquid interface.

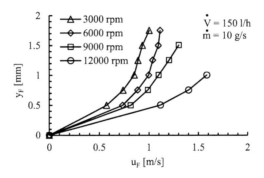

Fig. 2.5: Bearing chamber typical oil film velocity profiles (Glahn et al. (1996))

Based on the results, an attempt was made to find out whether the oil film flow in bearing chambers can be modelled/characterized by a theoretical approach which is required for a preliminary design in the development phase of bearing chambers. To achieve this, Glahn et al. (1996) assumed fully developed conditions and applied a force balance on a film fluid element. Substituting dimensionless quantities in wall coordinates ($u_{\tau,F} = \sqrt{\frac{\tau_w}{\rho_F}}$, $y_F^+ = y\frac{u_{\tau,F}}{\nu_F}$) in the force balance equation and rearranging results in an interface to wall shear stress ratio as a function of 'σ' and 'y_F^+'.

$$\tau_i = \tau_w - \rho_F g \sin \varphi^* y \Rightarrow \frac{\tau_i}{\tau_w} = 1 - \sigma y_F^+ \ with \ \sigma = \frac{\nu_F g \sin \varphi^*}{\left(\frac{\tau_w}{\rho_F}\right)^{\frac{3}{2}}} \tag{2.1}$$

The total shear stress in the film results from the superposition of the molecular as well as the turbulent shear stress. Using the wall coordinates, the interface to shear stress ratio is given by the equation

$$\tau_i = \rho_F u_{\tau,F}^2 (1+\frac{\nu_t}{\nu_F})\frac{du^+}{dy^+} \Rightarrow \frac{\tau_i}{\tau_w} = (1+\frac{\nu_t}{\nu_F})\frac{du^+}{dy^+} \tag{2.2}$$

Comparing Eq. 2.1 and Eq. 2.2, a universal velocity profile for the shear and gravity driven wall film was found.

$$1-\sigma y_F^+ = (1+\frac{\nu_t}{\nu_F})\frac{du_F^+}{dy_F^+} \tag{2.3}$$

The only unknown in Eq. 2.3 is the turbulent viscosity 'ν_t', which can either be approximated by a fixed value (e.g. in laminar case $\nu_t = 0$) or an experimentally driven empirical correlation. Hence, the semi-analytical expression developed by Glahn et al. (1996) can reproduce a laminar as well as a turbulent velocity profile. Glahn et al. (1996) showed that the accuracy of the analytical approach (Eq. 2.3) strongly depends on the definition of the turbulent viscosity. For the measured velocity profiles in the bearing chamber test rig (Fig. 2.5), the empirical correlation of Deissler (1954) given by Eq. 2.4, yielded a good match with the data points as can be seen in Fig. 2.6. Therefore, Glahn et al. (1996) concluded that the film flow in bearing chambers can be characterized as turbulent.

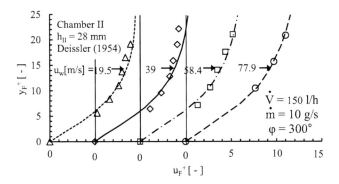

Fig. 2.6: Theoretical prediction of the film velocity profiles (Glahn et al. (1996))

$$\frac{\nu_t}{\nu_F} = m^2 u_f^+ y_f^+ [1-exp(-m^2 u_f^+ y_f^+)] \; with \; m = 0.109 \tag{2.4}$$

However, Deissler (1954) correlation for the turbulent viscosity can predict a laminar film flow as well because as the dimensionless film thickness 'y_f^+' in Eq. 2.4 tends to zero, the viscosity ratio also tends to zero. Therefore, the conclusion that the film is definitely turbulent because the Deissler correlation for eddy viscosity can reproduce the measured velocity profile is not valid. On the other hand, due to the chamber curvature (changing gravitational influence) and off-take disturbances, a continuous change in the film thickness is evident in Figs. 2.3 − 2.4 (Left); therefore, the film flow in bearing chambers especially 30° upstream the scavenge port cannot be considered as fully developed. Moreover, coexisting flow phenomena like the droplets hitting the film surface were neglected. However, the momentum transferred by the droplets to the film will result in an over or underestimate of the measured film velocity profile, hence, cannot be neglected. In fact, it can be theoretically and practically established that droplets hitting the film surface depending on the angle of incidence can enhance or reduce the film velocity. A point to be noted here is that Glahn et al. (1996) did not use seeding particles for the LDV – setup to measure the velocity profiles. Rather, the finely dispersed gas particles in the oil film served as the seeding particles. The presence of finely dispersed particles in the oil film can also be seen as an indirect indication of strong interaction with the core gas flow including air ingestion due to droplets hitting the film surface. Hence, the velocity measurements will be either over or underpredicted as it would be without droplets. Glahn (1995) did confirm qualitatively and quantitatively that a droplet rich core (gas) flow exists above the film surface for the investigated operating conditions. Therefore, it can now be stated that the film cannot be characterized as turbulent based on the velocity profile alone as the velocity profile might be different in the absence of droplet interaction. For a proper film characterization, the velocity profile must be determined either isolated from the droplets or the droplets interacting with the film at the location must be quantified.

Schmälzle (1997) used the test rig arrangement of Glahn et al. (1996) and measured the film velocity profiles near the scavenge off-take in the co- as well as the counter-current flow regime. A local film thickening followed by a recirculation region downstream of the scavenge port was found on the counter-current side (see Fig. 2.7). Due to the difficulties in measurements mentioned by Schmälzle (1997), only one shaft speed ($n = 4{,}000rpm$) was studied in the counter-current regime and it was assumed that a more uniform film distribution will occur for higher shaft speeds.

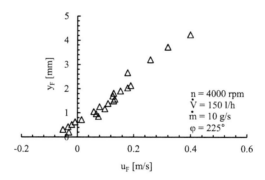

Fig. 2.7: Film recirculation on the counter-current side (Schmälzle (1997))

On the co-current side, Schmälzle (1997) measured the film velocity profiles $30°$ upstream of the scavenge off-take. From the measurements, it was obvious that an increase in the shaft velocity results in an increase in the film velocity due to the increase in the interfacial shear stresses. A significant increase in the film velocity, however, could only be recorded for $n \geq 8{,}000rpm$ which demonstrates the strong influence of wall (viscous stresses) on films less than or equal to a millimetre (Fig. 2.8 (left)).

Schmälzle (1997) also investigated the influence of oil feed on the film dynamics by testing three oil feeds ($50l/hr, 100l/hr$ & $150l/hr$) for a constant shaft speed of $8{,}000rpm$ (Fig. 2.9). For

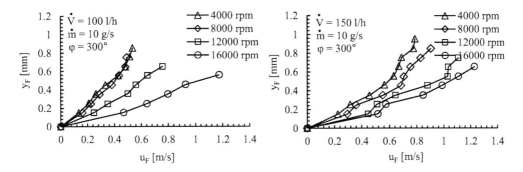

Fig. 2.8: Influence of shaft speed on the film velocity profile (Schmälzle (1997))
($100l/hr$ (Left); $150l/hr$ (Right))

increasing the oil feed, considerable increase in the film velocity and slight increase in the film thickness was recorded. The fact that the influence of gravitational and interfacial shear forces increase with film thickness, explains the considerable increase in the film velocity and only a slight increase in the film thickness. The measured velocity profiles were also compared with the theoretical approach introduced by Glahn et al. (1996) and a good agreement was found.

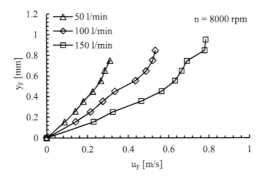

Fig. 2.9: Influence of oil feed on the film velocity profile (Schmälzle (1997))

Lipp (2003) provided extensive qualitative insights into the two-phase flow phenomena in a bearing chamber under aero engine relevant conditions. The focus was placed on the counter-current regime near the scavenge port and it was found that for a bigger chamber configuration ($b \times h = 44 \times 47mm$) , an up to $20-30mm$ thick recirculation region was present at all times (see Fig. 2.10). Only above $12,000rpm$, the recirculation region vanishes and the film becomes uniformly distributed along the chamber wall. Glahn (1995) also reported a similar trend in relatively smaller chambers where a shaft speed of at least $n \geq 12,000rpm$ was necessary to achieve a homogenously distributed film. The oil split between the vent and scavenge pipes was also measured for a large number of parameter variation.

Gorse et al. (2003) reiterated the strong dependence of oil film dynamics on the air flow inside the bearing chambers and pointed out the necessity of understanding the pure air flow field inside

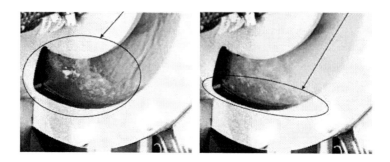

Fig. 2.10: Visualization of the near off-take film flows (Lipp (2003))

the bearing chambers as well. Accordingly, the model bearing chamber test rig was adapted to the new measurements and a specially adapted three dimensional Laser Doppler Anemometry (LDA) system was used to resolve the complex flow. The rotational speed and the sealing air flow were varied systematically in a wide range. The results revealed that the flow field strongly depended on the ratio of the shaft's surface velocity to the axial velocity at the inlet of the bearing chamber. If this ratio was greater than 4.2, two secondary vortices occurred in the chamber whereas below only one sealing air driven vortex was determined. Depending on the established mode, differences were also observed in the tangential velocity component. Nevertheless, the measured tangential velocity at the exit of the labyrinth seals was in good agreement with the former investigations e.g. Morrison et al. (1991) i.e. $20-30\%$ of the shaft speed.

Gorse et al. (2004) complemented the investigations carried out by Glahn (1995) by a new chamber geometry $(b \times h = 15 \times 10mm)$. Likewise, the influence of operating conditions (shaft speeds and oil flow rates) and chamber geometry on the circumferential oil film thickness distribution were investigated in the new geometry and compared to the data of Glahn (1995). Capacitive probes were used to measure the film thickness at five circumferential positions. Keeping the oil feed constant to $100l/hr$, a mean film thickness distribution in the range $0.5mm < \overline{h_f} < 1.2mm$ was found for all shaft speeds.

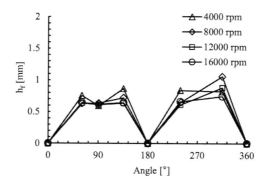

Fig. 2.11: Influence of shaft speed on the film thickness distribution (Gorse et al. (2004))

The effect of oil feed on the circumferential film distribution at two shaft speeds, $8,000rpm$ (see

Fig. 2.12 (left)) and $12,000rpm$ (see Fig. 2.12 (right)) respectively, was also studied. For all investigated oil feeds $(50l/hr, 100l/hr \ \& \ 150l/hr)$, Gorse (2006) recorded almost a similar trend both in magnitude and distribution as measured by Glahn (1995). The measurements showed that the film thickness ranged between $0.5mm < \overline{h}_f < 1.2mm$ for moderate shaft speeds ($8,000rpm$) and decreased to the range $0.5mm < \overline{h}_f < 1.0mm$ for higher shaft speeds ($12,000rpm$). From the experiments, once again it was evident that the operating conditions as well as the design of the chamber geometries strongly influenced the oil film dynamics. Moreover, it was found that the oil film was strongly affected by the design of the vent and scavenge ports, which also strongly influenced the discharge characteristics of the air/oil mixture.

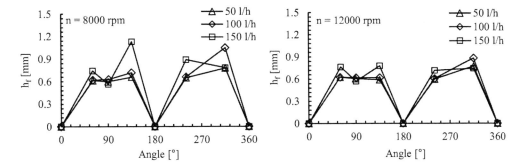

Fig. 2.12: Influence of oil feed on the film thickness distribution (Gorse et al. (2004)) ($8,000rpm$ (Left); $12,000rpm$ (Right))

A 1D analytical approach based on the previous works of Glahn (1995, 1996) was introduced to predict the circumferential film distribution in the chamber. This approach required the knowledge of film thickness and velocity profile at one angular position which were measured experimentally at an offset of $15°$ i.e. $135°$ and $150°$ respectively. The LDV technique, as explained by Glahn et al. (1996), was used to measure the film velocity profile. In addition to that, a theoretical velocity profile was also required which was calculated using Eq. 2.3 and fits well to the measured data as can be seen in Fig. 2.13.

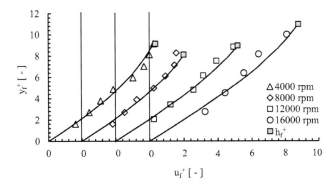

Fig. 2.13: Theoretical prediction of the film velocity profiles ($100l/hr$, Gorse et al. (2004))

Two assumptions, namely, a constant interfacial shear stress on the film surface and a linear increase in the film loading along the chamber housing, were also made. The tangential velocity, estimated using an angular momentum balance, was used to quantify the interfacial shear stress on the film surface. When compared with the experimental data, the angular momentum balance predicts the tangential velocity with a good accuracy and proves to be a good tool for comparing different chamber configurations. With the help of this simplified analytical approach, the authors aimed to devise a tool which can anticipate the behaviour of film flow in advance and can help to characterize the film flow in bearing chambers. However, as can be seen from Fig. 2.14, only limited success was achieved using this technique. The authors concluded their work with the comment that additional and more generic research is necessary to further improve the understanding of film flow phenomena in bearing chambers which will help in developing advanced analytical approaches and accurate design rules for future bearing chambers.

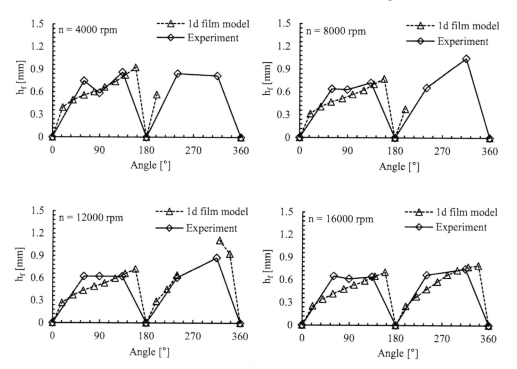

Fig. 2.14: 1D film model to predict the circumferential film distribution (Gorse et al. (2004))

From similar investigations outside ITS, the experiments conducted by Flouros (2005), Chandra (2006) and Chandra et al. (2010) are most relevant. Flouros (2005) conducted a parametric study on the bearing power consumption with a 124 mm pitch circle diameter ball bearing in a bearing chamber adapted from the RB199 turbofan engine. The test facility used by Flouros (2005) is shown in Fig. 2.15. Torque measurements were used to measure the power consumption of the bearing and to enable the comparisons between different operating conditions. The operating parameters such as oil flow, oil temperature, sealing air flow, bearing chamber pressure and shaft speed were varied to assess the impact on the power consumption. Besides parameter variation, Flouros (2005) also conducted a large number of parameter combination studies and found out that almost in every case, increasing the oil flow rate results in a continuous increase in power consumption. Although no direct information on the oil film dynamics, neither film thickness nor the velocity profile, was reported; yet, valuable information can be derived from the power consumption measurements as a function of oil flow rate supplied to the bearing chamber. This can easily be done based on the experience gained from the investigations conducted using the ITS model bearing chamber test rig where increasing the oil flow rate has a direct influence on the oil film thickness and velocity profile (e.g. Figs. 2.4, 2.8, 2.9, 2.12). Keeping this in mind, it can be generalized that increasing the oil flow rate results in an increase in the chamber resistance. For a constant shaft speed, this additional resistance has to be overcome by the drivetrain hence compensated by an additional torque at the shaft.

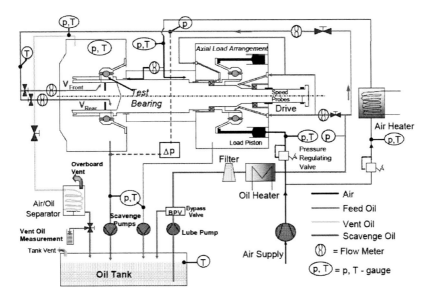

Fig. 2.15: Test facility to investigate bearing power consumption (Flouros (2005))

The test rig used by Chandra (2006) was provided with a rotating shaft and a provision to test different scavenge off-take designs (see Fig. 2.16). Liquid film enters the test rig concurrently with the rotating shaft's tangential velocity. The liquid flow rate, scavenge ratio and shaft rotational speed were varied and their influence on the sump residence volume and scavenge air/liquid discharge ratio was studied. Many interesting factors affecting the removal of oil through a generic scavenge system were identified. Nevertheless, it has to be mentioned at this point that most of these investigations were carried out under simplified flow conditions e.g. no pressurized air, labyrinth seal was absent and vent off-take was abstracted by a hole exposed to the atmospheric conditions in the test rig. Therefore, the transfer of knowledge to the real bearing chamber environment is difficult or even questionable.

In the most recent investigation at the ITS, Peduto et al. (2010) determined the circumferential

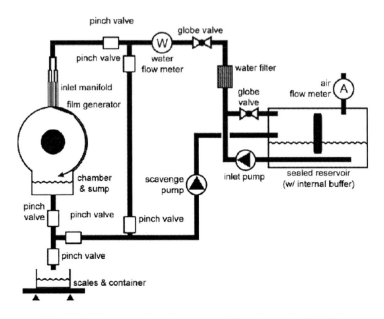

Fig. 2.16: Bearing chamber test rig (Chandra et al. (2010))

distribution of oil film at 10 angular positions located in the axial middle of the bearing chamber configuration similar to that of Glahn et al. (1996). The experiments were conducted at atmospheric as well as at elevated pressures up to $3bar$ (absolute). Effect of shaft speeds and oil feeds were studied and it was found that the film thickness at elevated pressures was much smaller and relatively uniformly distributed as compared to the measurements at atmospheric pressure. This behaviour is due to the fact that density is a function of pressure and the interface shear stress is directly proportional to the density, hence, film experiences higher shear even at lower shaft speeds. The measured film thickness distribution was comparable to that of Glahn (1995), Schmälzle (1997) and Gorse et al. (2004).

Analysing the comprehensive investigations in the ITS model bearing chamber test rig, Gorse (2007) was able to elaborate the film generation mechanism in the bearing chambers. According to him, the oil film observed in the model bearing chamber test rig originates directly from the bearing as a result of under-race lubrication. The film is also fed by the large droplets/ligaments, created during the interaction between the rolling elements and the bearing cage, which are catapulted towards the bearing chamber outer wall. From the film generation mechanisms, it can be said that the film velocity is negligibly small as compared to the air velocity. The air velocity in the bearing chambers can be estimated from the consideration that the fast rotating shaft accelerates the air to almost $20-30\%$ of the shaft's tangential velocity (Gorse et al. (2003), Morrison et al. (1991)). In a high pressure bearing chamber, shaft speeds up to $10,000-15,000rpm$ occur whereas a typical intermediate pressure bearing chamber observes $5,000-7,000rpm$ during engine operation. Assuming a general case where the shaft rotates with $10,000(6,000)rpm$ and has a typical diameter of $\varnothing 100mm$ then a mean air velocity of approx. $15m/s(10m/s)$ results in

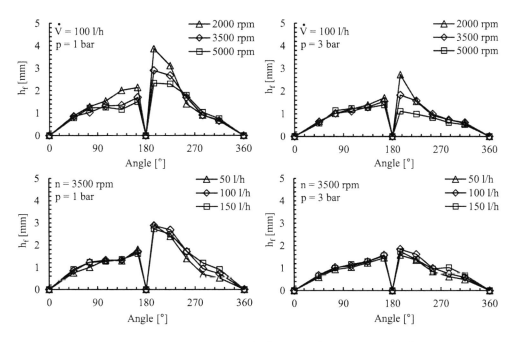

Fig. 2.17: Film thickness distribution in the HSBC test rig (Peduto et al. (2010))
        (At atmospheric pressure (Left); At high pressures (Right))

the high (intermediate) pressure bearing chamber. The air flow in the bearing chambers influences the dynamics of the oil film running on the chamber wall. The resulting behaviour of the oil film flow together with the coexisting flow phenomena which might be governing the real engine bearing chambers was qualitatively summarized by Gorse (2007) in the manner as seen in Fig. 2.18.

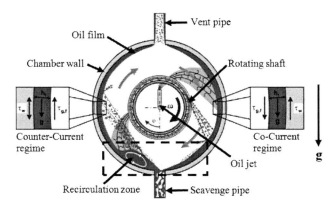

Fig. 2.18: Probable two-phase flow phenomena in bearing chambers

Based on the directions of the two most important driving forces, the shear and the gravitational forces, two flow regimes can be identified in Fig. 2.18. The first region has the driving forces in the same direction and called the co-current regime whereas the second region has the driving forces in the opposite directions and called the counter-current regime. It should be noted here that the term counter-current only indicates that the major forces influencing the film dynamics are acting in the opposite directions and has no indication about the direction of the film flow. The film may or may not follow the shearing gas flow direction.

Now from the film thickness measurements presented earlier, it can be said that except in the vicinity of the scavenge off-takes $(135-225°)$, the film thickness is usually less than a millimetre and can be considered to remain constant. Assuming that the film surface is exposed to an averaged shear force resulting from the air flow, the characteristic time-averaged parameters can be used for describing the film flow under the action of interfacial shear stresses. This treatment of the shear driven film flows refers to the thin film modelling approach. The model equations are 2D parabolic (the flow field is updated based on the upstream conditions only) and require a prior knowledge of time-averaged local film thickness and velocity profile at a location. Although, thin film models cannot account for a detailed interaction between the phases including the unsteady wave characteristics and droplet stripping as shown by e.g. Himmelsbach (1992) and Wittig et al. (1992); yet, due to the importance of thin film flows for many technical applications, a considerable amount of work has been carried out on several relevant topics. At the Institut für Thermische Strömungsmaschinen, mostly dealing with the aero engine combustors, the experimental and numerical investigations of thin films have a long history. Wurz (1976), Sill (1980), Sattelmayer et al. (1986), Samenfink et al. (1996), Rosskamp et al. (1997a) and Schober et al. (2003) provided detailed experimental findings on the structure and behaviour of shear driven thin liquid films. Based upon these studies, Himmelsbach et al. (1994), Gerendás et al. (1995), Rosskamp et al. (1997b, 1998) and Ebner et al. (2004) proposed numerical methods for the prediction of motion and heat transfer to thin wall films including the phase change phenomena. Therefore, it can be stated here that the thin film models are fairly well-developed and deliver good results if the underlying assumptions hold.

In the vicinity of the scavenge off-takes, however, the film thickness can grow even beyond five millimetres for some configurations and on the counter-current side can also manifest an evidence of a recirculation within the film (see Figs. 2.2, 2.7, 2.10 & 2.17). Unlike the thin films, the film thickness in this case is neither negligible as compared to the chamber height nor it can be considered as constant. The film surface becomes unstable for very low shearing gas velocities and displays unique waveforms. The droplet generation mechanism also behaves differently and happens at much lower shearing gas velocities than required for thin films. Hence, the thick films behave completely different when compared to the thin films. Therefore, the assumptions of parabolic instead of elliptic equation set and fully developed flow conditions are inappropriate here. Accordingly, all theoretical treatments with such assumptions could neither be validated nor generalized for thick films. For example, the theoretical approach of Glahn et al. (1996) yields a good agreement with the measured velocity profile; however, the wall shear stress used to calculate the velocity profile gives a deviation of up to 40% from the measured value (Elsäßer (1998)). This means that the calculated velocity profile following the approach of Glahn et al. (1996) when extended to the wall will give a gradient with up to 40% error. As pressure drop predictions in the bearing chambers is one of the major objectives of researchers in this area,

40% error in the wall shear stress is unsatisfactory. Especially, the presence of recirculation in the counter-current regime totally contradicts the basic assumption of thin film models which do not support negative film velocities. However, not much attention has been given to the counter-current flow regime in the bearing chamber investigations. In the past, it was assumed that after a certain shaft speed ($n \geq 12,000 rpm$), only the co-current regime exists in the bearing chambers and the wall film is uniformly distributed along the circumference. This is a characteristic of rimming flows where the wall at which the film is attached rotates with high speeds. As this is not the case in bearing chambers and the fact that different air flow patterns can exist for different chamber geometries (Gorse et al. (2003, 2004)), a clean rimming flow cannot be assumed for all chamber configurations (see Figs. 2.3 - 2.4 (left), Fig. 2.10) or at least for transient engine behaviour e.g. abrupt engine shutdown. Furthermore, the aero engine manufacturers experience the problem of coking which is a direct result of increase in the residence time of the oil film in the bearing chambers. The recirculation zone on the counter-current side is one of the modes of increase in the residence time. Therefore, it is necessary to investigate the thick film dynamics in the co- as well the counter-current regimes.

On the other hand, due to the limitations posed by the thin film models, the thick films cannot be modelled using simplified analytical approaches. Consequently, there is a growing interest in analysing the potential offered by the full 3D multiphase numerical techniques. However, for an effective and reliable analysis of the multiphase models well-defined boundary conditions are necessary which are either vague or unknown in the complex bearing chamber test rigs. Therefore, to fulfil the requirements identified here, it is necessary to think of ways where such investigations of the thick films can be performed in isolation. As mentioned earlier, it is impossible to isolate the film flow from the coexisting flow phenomena e.g. droplet interaction, off-take disturbances etc...in the complex bearing chamber test rigs e.g. ITS HSBC. At this point, an idea of a simple test rig performing such investigations and delivering all the desired quantities can be envisaged by analysing the forces acting on an infinitesimal film element in the vicinity of the scavenge off-takes. As shown in Fig. 2.19, if the resultant '$\vec{F}_R$' of the gravitational '$\vec{F}_G$' and centrifugal '$\vec{F}_C$' forces makes an angle 'φ^*' with the horizontal then a linear rotatable test rig placed at some angle 'θ' to the horizontal will represent the angular position 'φ' of the film element according to the relationship given by Eq. 2.5.

$$\left(\frac{|\vec{F}_C|}{|\vec{F}_G|} + 1 \right) \sin \left(\frac{1}{\frac{|\vec{F}_C|}{|\vec{F}_G|} + 1} \varphi \right) = \sin \theta \Rightarrow \left| \begin{array}{l} \rightarrow \varphi = \theta \\ \rightarrow 1 \leq \varphi \leq 28.96° \approx 1 \leq \theta \leq 30° \\ \rightarrow 1 \leq \varphi \leq 28.78° \approx 1 \leq \theta \leq 30° \\ \quad for \frac{|\vec{F}_C|}{|\vec{F}_G|} = 0, 1 \ and \ 2 \end{array} \right. \quad (2.5)$$

Therefore, the focus of the experimental work presented in this thesis is set to utilize a linear rotatable test rig to perform the required investigations of the shear driven thick wall films similar to the oil film in the vicinity of the scavenge off-takes. However, before proceeding to the materialization of this idea, it is necessary to go through the research activities of similar nature in other engineering fields. In this way, it can be made sure that the investigations in the framework of this research work are unique and fill up an important gap in the present scientific literature.

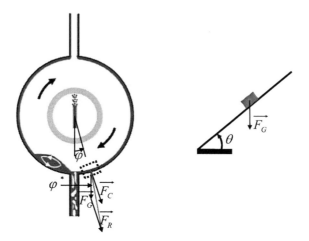

Fig. 2.19: Transformation of the near off-take films to a linear rig

In the following, typical test rig designs and the investigations carried out in this respect are summarized.

2.1.2 Stratified Flow Test Rigs

For a simultaneous gas-liquid flow, the stratified flow pattern is considered to be among the simplest; however, it is far from being completely understood. The lack of details is mainly due to the wavy nature of the interface and the interaction between the deformed interface and the flow structure in both phases (Berthelsen and Ytrehus (2005)). The interaction between turbulent gas flows and moving liquid streams has been a subject of intense research in the last several decades, mainly due to the importance of transport processes at the gas-liquid interfaces in the chemical/petrochemical industries, heat exchanger ducts, nuclear reactors and geophysical sciences. The local structure of turbulence at moving gas-liquid interfaces is an important factor in the transport of mass, momentum and energy between the gas and liquid phases in most two-phase flows. An understanding of the mechanisms which create, convect and diffuse turbulence at moving interfaces is also essential as a basis for more accurate mathematical modelling of the interfacial transport processes (Lorencez et al. (1997), Lee(1992)). Numerous experimental studies of local flow properties have been devoted to stratified wavy gas-liquid flow, but difficulty in making accurate measurements close to the interface limits the insight to this flow regime (Berthelsen et al. (2005)). However, the requirements for economic design, optimization of operating conditions and assessment of safety factors create the need for quantitative information about such flows (Ghorai et al. (2005)). Therefore, a large number of experimental efforts have been made in the past to get reliable estimates of the hydrodynamic properties, mostly pressure gradient and hold-up e.g. Badie et al. (2000), Ottens et al. (2001), Ferreira (2004) etc..., associated with this type of flows. The most common experimental setup consists of a long pipe or rectangular/square duct and deals with fully developed flows. The length of the test section usually amounts to a couple of meters ($\sim 8m$, Vallee et al. (2006)), however, to ensure fully developed conditions others have suggested much longer test sections e.g. Lopez (1994)/Strand (1993) used a $25m/35m$ long test section. The investigated flow variables include pressure drop, liquid holdup/void fraction, film thickness, velocity distribution, surface wave classification in the form of flow maps. Numerous authors involved and their work, in this area of research, can be named which dates back to as early as the 1900's (Cornish (1910), Jeffreys (1925)). Some of the relevant investigations are reviewed in the following sections.

2.1.2.1 Co-Current Flow Regime

Investigations involving the qualitative behaviour of a gas-liquid interface with variable film properties

Hanratty and Engen (1957) investigated interactions between a turbulent air stream and a moving water surface using approx. a $6m$ long horizontal channel made of Plexiglas. The data were correlated in terms of liquid and gas Reynolds numbers. The concept of critical lower limit on the gas velocity below which the water surface remains flat was introduced. The effect of the gas flow on the liquid was described in terms of five types of liquid surfaces:(1) smooth, (2) two-dimensional waves, (3) squalls, (4) roll waves, and (5) dispersed flow. It was reported that the transition from a smooth surface to a surface possessing 2D waves occurred at about $Re_L \sim 520$ whereas the transition from 2D waves to a "pebbled" surface occurred at about $Re_L \sim 600$. Thinner films or films having a lower Reynolds number were found to be more stable. It was mentioned that theories presented in the literature for the initiation of waves on a liquid surface did not adequately describe the experimental results.

Hanratty and Hershman (1961) carried out experiments in the same equipment designed by Hanratty and Engen (1957) to study the roll-wave transition for a horizontal air-liquid flow. The liquid viscosity was varied in the range $1mPa.s \leq \mu \leq 6.60mPa.s$ and surface tension in the range $35mN/m \leq \sigma \leq 75mN/m$. This was accomplished by using glycerine water solutions and butanol-water (sodium lauryl sulphate) solutions. Roll-wave transition was determined at fixed liquid rates by varying the gas flow rate. The roll waves appeared to form very close to the entry, and they covered very nearly the whole width of the channel. Changes in the design of the liquid entry did not affect the transition conditions. No significant difference was noted in the surface structure between the runs with water (high surface tension) and the runs with butanol solutions (low surface tension). However, in the runs with solutions of sodium lauryl sulfate (low surface tension) the surface was completely smooth and the flow was determined to be laminar by tracer studies at the roll-wave transition. Hence, a contradictory statement was delivered on the influence of surface tension on the free surface instability. This is only possible if the employed surfactant altered other physical properties of the solution too.

Cohen and Hanratty (1965) again explained the formation of 2D and 3D waves based on the concept of critical gas velocity. The transition from a 2D to 3D wave regime was determined visually and by the inspection of the pressure drop data. The definition of 3D transition was more arbitrary than the definition of the 2D transition. The 3D waves were said to occur when the characteristic lengths of the waves in the direction of flow becomes equal to the characteristic length perpendicular to the direction of flow. Like Hanratty & Engen (1957) they also observed that the critical gas velocities increased with a decrease in the liquid height and an increase in the liquid viscosity. To further support their finding they mentioned Hershman who investigated a range of fluid viscosities from $0.8267mPa.s \leq \mu \leq 13.2267mPa.s$ (water has a viscosity of approx. $0.8989mPa.s$) and also found the same behaviour i.e. the critical gas velocity increased monotonically with decreasing liquid Reynolds number. The two dimensional waves were found to be more stable the more viscous the liquid was. For water, 3D waves appeared at only slightly higher gas velocities than were needed to generate 2D waves. For the more viscous liquids, the 2D waves were stable over a large range of gas flow rates.

Weisman et al. (1979) obtained extensive data on the transitions between two-phase flow patterns during a co-current gas-liquid flow in horizontal pipes. Fluid properties were varied in a systematic manner to determine the effects of liquid viscosity, liquid density, surface tension and gas density. Pipe diameters were also varied from $12-50mm$ for most of the tests. Visual observations were supplemented by an analysis of pressure drop fluctuations and hence, the data was believed to be less subjected than most previous observations. The transition boundaries were compared with the air-water flow map. It was observed that high viscosity had no considerable influence on the smooth to wavy stratified transition. On the other hand, decreasing the surface tension resulted in the wavy stratified transition from a smooth surface at significantly higher gas flow rates. Increasing the density also resulted in a late transition from stratified to wavy flow i.e. shifted to higher gas flow rates. However, it should be noted at this point that the method adopted for increasing density also altered the viscosity (slightly) and surface tension (almost by 33%); therefore, the observation regarding increased density cannot be considered as fully independent.

Andritsos and Hanratty (1986) carried out experiments in a horizontal circular pipe with two diameters and characterize the different type of wave patterns observed for the air-liquid flow. The liquid viscosity was varied in the range $1mPa.s \leq \mu \leq 80mPa.s$. It was reported that the high viscosity wave patterns differ from those of low viscosity liquid in that, the region of regular 2D waves barely exists and the interface appears less roughened even when waves were present. Hence, it was suggested that viscosity has a dampening effect on the surface instabilities.

Andritsos and Hanratty (1987) continuing their work (Andritsos and Hanratty (1986)), obtained measurements of liquid height and pressure drop for fully developed horizontal pipe flows. They found that the interfacial friction factor 'f_i' increases linearly with gas velocity at gas velocities larger than needed to initiate waves and that the proportionality constant is insensitive to the pipe diameters. The effects of liquid viscosity and liquid flow rate were found to be of secondary importance over the range of liquid viscosities studied and were taken into account by assuming that 'f_i' is a function of the ratio of the height of the liquid to the pipe diameter 'h/D'.

Hart et al. (1989) used a straight horizontal copper tube with $\varnothing 51mm$ inner diameter and length approx. $17m$ including a $1.4m$ precision glass section. The glass tube was about a distance of $4m$ from the liquid injection point and was provided with an angle gauge to measure the wetting angle made by the liquid. Measurements were carried out at superficial air velocities ranging from $5-30m/s$ and superficial liquid velocities ranging from $0.00025-0.08m/s$ which correspond to film thicknesses ranging from $\sim 0-3mm$. Flow regimes observed during the experiments were stratified, wavy stratified and annular, as has been indicated in the flow pattern map of Mandhane et al. (1974). The effect of surface tension ($38mN/m \leq \sigma \leq 75mN/m$) was investigated by using water in combination with a surface active agent 'Tween 80' and viscosity $(0.9mPa.s \leq \mu \leq 8.5mPa.s)$ by using water in combination with glycerine. It was shown that in the test range, surface tension and viscosity of the gas-liquid system has neither noticeable effect on the liquid hold-up (film thickness) nor on the wetting wall fraction. The pressure gradient, however, increases slightly with decreasing surface tension values and was explained on the basis of open literature (Friedel (1977)) i.e. a decrease in the surface tension promotes formation and magnification of waves hence surface roughness. The pressure gradient was also shown to be a function of the liquid hold-up.

Investigations involving the qualitative behaviour of a gas-liquid interface leading to flow maps
Due to the importance of flow regime knowledge in advance, most authors dealing with the different waveforms appearing on a gas-liquid interface have summarized the transition boundaries, (most commonly) as a function of superficial gas and liquid velocities, in flow maps. The flow maps were usually compiled on the basis of qualitative observations either made using naked eyes or techniques like high speed imaging (e.g. Baker (1954), Mandhane et al. (1974), Taitel and Dukler (1976), Shoham (1984), Lin & Hanratty (1987), Petalas & Aziz (1998) etc...), however, few quantitative attempts using e.g. PDF and PSD of pressure fluctuations (Bruno & McCready (1989), Strand (1993), Shi & Kocamustafaorgullari (1994), Shim & Jo (2000), Fernandino & Ytrehus (2006)) have also been made.

Investigations involving the velocity and film thickness (hold-up) measurements
Paras and Karabelas (1992) measured film axial velocity components in a $16m$ long horizontal pipe using LDA. The data obtained suggested that only in the vicinity of the solid surface (sublayer) does the liquid film motion resemble the well-known behaviour of single phase flow. Beyond that, the flow field is strongly influenced by the wavy gas-liquid interface and by the apparently intensive energy transfer from the very fast moving gas to liquid layer.
Dykhno et al. (1994), Flores et al. (1995) and Paras et al. (1998) have conducted experiments in $26, 3$ and $16m$ long circular pipes and determined the gas phase velocity field.
Paras et al. (1994) & Vlachos et al. (1997) have made measurements of local velocities and wall shear stress in the liquid phase. A $5m$ long pipe was used to conduct the experiments.
Fabre et al. (1987) carried out measurements of film thickness and velocity distribution in both (liquid and gas) phases using a $13m$ long rectangular section. In general, it can be said that the velocity and film thickness data found in literature was determined for very long pipes/channels to ensure fully developed conditions.

Investigations involving turbulent quantities
The nature of turbulence in wavy stratified flows in a horizontal channel has been studied by a number of investigators such as Johns et al. (1975), Akai et al. (1981), Fabre et al. (1983, 1987), Murata el al. (1991) and Lorencez el al. (1991, 1993) among others, however, the understanding of the turbulence structures near a sheared and wavy interface is not yet fully satisfactory (Lorencez et al. (1997)).
Rashidi and Banergee (1990) reported that at high interfacial shear stresses, turbulent behaviour near the gas-liquid interface is comparable to the observations made at the liquidwall interface i.e. turbulent streaks appearing at the smooth interface broke down to form bursts.
Continuing his work, Lorencez et al. (1997) studied the interfacial momentum transfer in a

rectangular channel using HWA and non-intrusive photochromatic dye activation (PDA) techniques. The wavy interface was found to change the kinematic and turbulent structure of the liquid phase significantly, resulting in nearly constant streamwise velocity in the upper half of the liquid phase. The streamwise velocity fluctuations were observed to be enhanced primarily by the intense shearing in the upper layers of the liquid, while the increase in magnitude of the vertical fluctuations was associated mostly with the increase in interfacial wave amplitude. The strong interaction between the gas flow and the waves produced interfacial turbulent bursts that considerably increased the mixing in the upper layers of the liquid. The magnitudes of the streamwise and vertical turbulent fluctuations were correlated to the interfacial shear and wave amplitudes.

2.1.2.2   Counter-Current Flow Regime

Counter-current two-phase flows can be encountered in a variety of industrial applications, such as wetted wall columns and tubular contactors as well as in the safety analysis of nuclear reactors. In the case of a counter-current gas-liquid flow, a thickening of the film is known to occur (Feind (1960)). Upon increase of the relative velocity between the liquid film and the gas beyond a 'critical' value, the film may be disrupted and part of the liquid sheared off the surface wave crests and entrained as droplets in the gas stream (Steen and Wallis (1964), Arnold and Hewitt (1967)). At even higher gas flow rates, a condition called *flooding* at which the net downward liquid flow rate becomes zero may occur (Cetinbudaklar and Jameson (1969), Imura et al. (1977), Lee and Bankoff (1982)). Increasing the gas flow rate beyond this point results in a complete flow reversal. In the safety analysis of nuclear reactors, understanding of counter-current flow limitation (CCFL) phenomena is very important, particularly during a hypothetical loss-of-coolant accident (LOCA). However, the physical mechanisms responsible for the onset of flooding have not been perfectly understood despite over fifty years of research. Most of the previous experimental studies have focused on the measurements of liquid film thickness and surface wave characteristics.

Investigations involving the qualitative behaviour of a gas-liquid interface with variable film properties
Ghiaasiaan et al. (1997) experimentally investigated the flow patterns, flow limitation (flooding), and gas hold-up (void fraction) in counter-current flow in vertical and inclined pipes. Tests were performed in a $2m$ long pipe with $19mm$ inner diameter, using air, and demineralised water, mineral and paraffin oils, covering a surface tension range of $12.8mN/m \leq \sigma \leq 72mN/m$, and a liquid viscosity range of $1mPa.s \leq \mu \leq 185mPa.s$. The liquid and gas superficial velocities for the tests with demineralised water were in the range $0-0.54m/s$ and $0.01-2.99m/s$, for tests with mineral oil were in the range $0-0.23m/s$ and $0.01-0.248m/s$, and for paraffin oil were in the range $0.0015-0.116m/s$ and $0.01-2.24m/s$, respectively. The examined pipe angles of inclination with respect to the vertical line were $0°$, $30°$ and $68°$. Flooding data were significantly different from pure water results only at very high liquid viscosities. The effect of liquid viscosity on gas hold-up and flow patterns was significant. With increasing the liquid viscosity the parameter range of the slug flow pattern expanded for all angles of inclination, and froth flow replaced the churn flow pattern in the vertical configuration and it replaced the

churn/stratified and semistratified patterns in inclined configurations. The churn-stratified flow pattern was reported to be predominantly wavy stratified and was interrupted by upward moving flooding type waves. Semi-stratified was a periodic pattern where in each period, the flow regime was initially wavy stratified while liquid accumulated in the bottom portion of the test section and formed a large liquid slug which subsequently moved upwards in the channel.

Investigations involving the qualitative behaviour of a gas-liquid interface leading to flow maps
Yamaguchi and Yamazaki (1982, 1984) performed experiments in vertical pipes with 40 and 80mm inner diameters, and identified four major flow regimes: bubbly, which could be maintained only at very low superficial gas velocities; slug, which occupied an extensive portion of the entire flow map; semi-annular, which represented the transition between slug and annular regimes; and annular.

Zabaras and Dukler (1988) and Biage et al. (1989) observed that the surface waves on the liquid film propagate in the downwards direction but they were never large enough in amplitude to bridge the tube. Govan et al. (1991), on the other hand, claimed that flooding is caused by the upward propagation of a wave and the accumulation of other waves falling onto the initial one.

Ghiaasiaan et al. (1995a) performed experiments with air and water, using vertical and inclined channels. Flow patterns were studied and average gas hold-ups were measured using quick-action valves. The flow patterns, which were relatively sensitive to the channel angle of inclination, included some flow patterns which may not occur in co-current flows.

Gargallo et al. (2005) investigated the counter-current flow regime of air and water using the WENKA test facility at the Forschungszentrum Karlsruhe. It was found that flow reversal will not occur as long as the film remains supercritical ($Fr > 1$). The visual observation always showed that the film underwent a hydraulic jump before the flow reversal occurs.

Investigations involving velocity field, film thickness & turbulent quantities
Roy and Jain (1989) studied the characteristics of thin water film flow down an inclined plane without as well as with superposed counter-current air flow. The results showed that the introduction of counter-current air flow over the falling film increased the mean film thickness and generated finite amplitude disturbances on the film surface. The frequencies appeared to be primarily dependent on the air velocity. The data also indicated that there was a decrease in the propagation velocity (in the direction of film flow) of these interfacial disturbances with increase in the air velocity. At high enough air velocities, large amplitude disturbances 'mini-slugs' formed on the film surface. Water eventually got torn off the crests of these and became entrained in the air stream as droplets. The observed structures on the water film surface compared well with the observations of several earlier investigators.

Lorencez et al. (1997), similar to the co-current measurements (Lorencez et al. (1997)), provided the measurements of all the turbulent parameters for counter-current regime as well.

Vijayan et al. (2001) conducted experiments in tubes of 25, 67 and 99mm diameter. Measurements of the pressure gradient, film thickness and down flow rate were made for a range of air and water flow rates under pre and post flooding conditions. The data showed that the pressure gradient did not increase appreciably until just before the onset of flooding and the mean film thickness under pre-flooding conditions can be well predicted by the falling film correlations.

However, post-flooding both quantities showed a significant deviation from the pre-flooding pattern.

Stäbler et al. (2005, 2006) continuing the work of Gargallo et al. (2005), using the fluorescent-particle image velocimetry (PIV) and a single tip resistivity probe measurement techniques, measured the distribution of time-averaged local velocity fields, time-averaged local velocity fluctuations and void fraction in liquid film up to the free surface. Velocities in the wall normal direction showed very low values compared to the streamwise velocities. Near the free surface, the liquid is decelerated by the counter-current air flow. The velocity fluctuations in vertical and horizontal directions increased remarkably within the wavy two-phase region.

2.1.2.3 Summary of Experimental Investigations

At the end of section 2.1.1, it was emphasized that more generalized investigations are necessary before the film dynamics in bearing chambers can be fully understood. Although, due to the importance of film flow phenomena in a large number of industrial applications, it is studied for a long time but it appears that there is a lack of fundamental studies, especially those dealing with low liquid hold-ups $(2mm \leq h_f \leq 5mm)$. Most of the experimental data found in literature is case dependent and lacks generality. Moreover, either very little is reported on the experimental boundary conditions or not much attention is given to this issue. This probably results from the fact that almost all of the mentioned experimental investigations were simulating fully developed conditions and therefore, conducted in very long channels/pipes, hence, the influence of boundary conditions was expected to fade out. However, as a consequence of these ambiguous boundary conditions, for any (semi) analytical treatment of the data, assumptions need to be made for the boundary conditions. This fact further deteriorates the already unsatisfactory quality of correlations predicting the two-phase pressure drop and liquid hold-up (or film thickness) and this is the reason why no single correlation has proven to be accurate even for comparable data sets of different authors. Many independent studies (e.g. Battara et al. (1985), Gregory and Fogarasi (1985), Tribbe and Müller-Steinhagen (2000), Pawloski et al. (2004), Revellin (2005), Ribeiro et al. (2006)) on the general applicability of these correlations report a deviation of $\pm 20\%$ in the best cases with as high as $\pm 25 - 40\%$ deviation generally reported. Furthermore, existence of fully developed conditions in a bearing chamber environment is very unlikely. Therefore, the influence of boundary conditions on the developing flow field cannot be expected to fade out quickly. The growing interest in numerical modelling (CFD) also requires well-defined boundary conditions for setting up a reliable simulation because the numerical calculations are usually very sensitive to the choice of boundary conditions. The major problem arises while defining the outlet boundary condition. For fully developed flows, a single outlet can be defined for both phases because of the fact that the phase split remains constant. In contrary to the fully developed conditions, defining a single outlet for undeveloped flow conditions, is neither feasible nor practical. This is because of the fact that the phase split is continuously changing, hence, a single outlet will require extensive information on the dynamic behaviour of film flow which is usually unknown and desired as output. Keeping in mind the undeveloped flow conditions, another challenge arises from the typically reported validation data. A velocity profile, pressure gradient or film thickness can be used as a sufficient validation data for fully developed conditions which is insufficient for undeveloped conditions where the flow field continuously changes in space and time. Appreciating the fact that multiphase modelling techniques are still in a development phase and need extensive validation before a suitable film model can be standardized, it can be stated that with the data in hand, the modern CFD technique cannot be thoroughly analysed. Therefore, at this point in this thesis, in the light of arguments made, an absolute need for the fundamental investigations providing well-defined and reliable boundary conditions and appropriate validation data on the bearing chamber typical thick film flows can be strongly endorsed.

2.2 Numerical Investigations

2.2.1 Film Flow Modelling in Bearing Chamber (Similar) Geometries

The highly dynamic oil film can bypass the off-takes (carry over effect) several times due to the inertia provided by the fast moving air flow. The increase in residence time can quickly result in oil coking. Heat transfer measurements in bearing chambers are very difficult to conduct under real engine conditions; therefore, correlations describing heat transfer to oil film cannot be standardized till today. Moreover, the mechanism of film coking needs additional information on the film thickness distribution and velocity which are unknowns for most engine operating conditions. Therefore, it is very difficult to formulate a reliable heat transfer correlation (Zimmermann et al. (1991)). Zimmermann et al. (1991) was the first to introduce the two-phase flow correlations into aero engine air/oil system calculations and particularly for bearing chambers. The lack of experimental data directly applicable to aero engines was compensated to some extent by in-house experiments but largely by using the data produced by the nuclear industry. Correlations for the two-phase pressure drop in different aero engine components were presented and a reasonable agreement with data was achieved. Despite the success, the author took a serious note of all the limitations in achieving engine relevant data hence development of generic correlations. The author concluded that for understanding the two-phase flow phenomena in aero engine components, numerical methods would likely to become an important tool in the near future. Realizing this fact, parallel to the experimental investigations, attempts to resolve the bearing chamber flow phenomena numerically have also been made in the past and can be found in literature. However, due to the complexity of bearing chamber flow phenomena, demands on necessary computational resources and the fact that the multiphase numerical techniques are still in the development phase, the numerical investigations are often restricted to highly abstracted test cases. The oil film flow in bearing chambers has also become a topic of several recent and past publications some of which are reviewed in the following.
Chew (1996) developed a simple 2D numerical model to study the oil film behaviour on the housing of a bearing chamber and the results were compared with the experimental data of Wittig et al. (1994). Analysis included the effects of surface friction, gravity and swirl of the oil at inlet. The model was expected to resolve films at high rotational speeds where 'rimming' dominates. The model gave an indication of the likely occurrence of this regime and predicted heat transfer rates, film velocities and film thicknesses with some good qualitative and quantitative agreement with measurements but significant discrepancies and uncertainties remain. The lack of a good

agreement between the calculated and measured heat transfer along the wall housing of the bigger chamber was attributed to a different (not rimming) flow regime incompatible with the model and absence of off-take (scavenge) modelling. The author concluded that while the present modelling approach may be useful, and could be developed further, computational fluid dynamics methods (CFD) appear to offer the best prospect for a more general model.

Wang et al. (2000) studied the angular motion of an oil film on the rotating shaft's surface of a typical aero engine bearing chamber geometry. It was shown that at low shaft speeds, the motion of the oil film was influenced by the gravitational and shear forces. However, at high enough shaft speeds, the influence of gravitational force was no longer significant and the oil film rotated continuously with the shaft. The study, however, was only supported by a qualitative analysis and lacked a substantial quantitative analysis.

In a parallel study, Farrall et al. (2000) developed an isothermal model for describing a 2D continuous, shear-driven film in an annular geometry. The film model was coupled with the single-phase CFD solution via the interfacial shear stresses. Preliminary work to include the heat transfer effects was presented in Farrall (2001). Uniform profiles for temperature and velocity were assumed to predict the average film temperature profiles at different angular locations. This approach was further extended by Farrall et al. (2003) to allow for the prediction of complex two-phase air/oil flows in aero engine bearing chambers. CFX-4.3 was used to calculate the air flow and trajectories of oil droplet in the chamber, whilst the oil film motion and interaction of oil droplets with the film were described by in-house sub-models. The calculated film thickness distribution showed fair qualitative agreement with the experimental data of Gorse et al. (2003). Further extensions to the model (Farrall et al.(2004)) included the addition of boundary conditions to account for the effect of geometric features like vent and scavenge ports. It was found that the predicted film thickness and associated bulk velocity were sensitive to the choice of boundary conditions applied at the vent and scavenge outlets.

Maqableh et al. (2003) used the Eulerian modelling approach to study the flow and heat transfer behaviour of turbulent two-phase flows in a 2D annular geometry. Air, oil film and oil droplets were assumed as three components and modelled separately. Oil droplets in the core air flow were modelled as a dispersed phase and a particle drag model was used to calculate the interphase momentum transfer. Air and liquid oil were modelled as continuous phases and a mixture model was used to calculate the interphase momentum transfer. The results indicated that the solution variables and flow field were highly sensitive to the choice of input parameters e.g. droplet diameter range and boundary conditions. Hence, such modelling techniques can yield plausible results but require more input data.

Shimo et al. (2005) developed separate (thin) film models for the oil film simulations on the seal runner and sump wall. Parametric studies were conducted using both seal runner and wall film models to identify the influence of engine speed, oil flow rate and sump geometry on the film characteristics. From the results, generalized expressions were developed for windage stresses on the wall film. The investigations also revealed several interesting phenomena including the fact that the oil film on the sump outer wall cannot be considered as a steady flow. The film was generally unsteady due to the influence of gravity, windage and droplet momentum forces.

Young and Chew (2005) investigated a number of test cases in the context of aero engine oil system applications using the Volume-Of-Fluid modelling approach. The application cases were simple but smartly selected to represent the class of flows encountered in aero engine oil systems. In each case, plausible solutions were obtained using the standard solution strategies in a commercial CFD code. In the context of aero engine oil system and bearing chamber flows, especially wall film flows, the authors concluded that the application of VOF approach has no insurmountable issues.

Flouros (2008) evaluated the analytical and numerical models for the heat transfer in the vent and scavenge lines of a real engine. The Eulerian model available in a commercial CFD code was employed to simulate the flow phenomena. The vent and scavenge temperatures were compared with the measurements and empirical correlation of Busam (2004) where a good agreement was found. The possible flow regimes based on the flow maps corresponding to the vent and scavenge lines were also compared to the simulation results. However, the interface was not traced; therefore, the film flow cannot be interpreted by this methodology.

Robinson et al. (2008) in his work focused on the possible flow regimes at the off-takes of an aero engine bearing chamber. In the first test case, the scavenge off-take flow was abstracted by draining of a liquid from a vessel. Building on the first test case, a 3D CFD investigation of a double intake scavenge system was conducted. The interaction between the phases, which would have been the case in a real scavenge system, was not taken into account. The standard Volume-Of-Fluid method available in a commercial CFD code was used to simulate the test cases. The results of the first test case showed that naturally occurring features such as discharge and surface structures can be accounted for with VOF. However, in the more complex dual inlet test case, the CFD predictions could only be verified qualitatively.

Robinson et al. (2010), as a continuation of his work on the modelling of off-take flows, simulated the draining of a wall film, moving on a straight plate, through a hole. The interaction between the phases (gas-liquid) was minimized by setting the gas velocity equal to the film velocity at the inlet. The effect of film loading and inclination angle on the film thickness distribution along the plate and free surface in the drainage pipe was compared with the in-house experimental data. To some extent, a qualitative agreement could be found in all cases but a fair quantitative agreement could only be found for the simplest test case. Sensitivity to boundary conditions was anticipated as the reason for the lack of agreement between the simulation and experimental results. However, as will be shown later in this work that most of these discrepancies can be attributed directly to the ill-treatment of interphasic interactions.

From the literature regarding the numerical modelling of an oil film in the bearing chambers, it can be said that the application of advanced CFD technique to bearing chamber typical film flows is still in its early phase. Especially the interphasic interaction is either ill-treated or not dealt at all, which is the major feature of wall films typical to bearing chambers. To gain some insights into the wall film modelling with interphasic interactions, analogies from other engineering fields were studied. It was found that similar to the experimental investigations, as discussed in section 2.1.2, a number of numerical investigations also exist in a wide variety of areas including the petroleum and chemical processing industries, in steam generation and refrigeration equipment and in nuclear reactors. In these industries, the prediction of pressure loss and void fraction in gas-liquid stratified flows has been of considerable research interest since the 1930's. In the following, a brief overview of the methodology to study the stratified flows with interphasic interactions is outlined.

2.2.2 Stratified Flow Modelling

As mentioned in section 2.1.2, to achieve an economic design, optimization of operating conditions and assessment of safety factors, a prior knowledge on flow field especially pressure drop and hold-up is necessary. Consequently, the wavy stratified flows need to be modelled. In the past, this is mostly done either using generalized empirical data correlations or simple mechanistic models. Till today, no empirical correlation yielded satisfactory results against vast experimental data (see section 2.1.3). More recently, attentions has turned to the so-called mechanistic models to formulate semi-analytical calculation methods based on a greatly simplified representation of the flow structures. Mechanistic models, as reviewed by Berthelsen and Ytrehus (2005), are applied with common assumptions that both phases are treated as one-dimensional bulk flows. The shear stresses are calculated from empirical correlations based on the average velocity. Early approaches used modified single phase correlations, such as the well-known Blasius formula (e.g. Agrawal et al. (1973), Russel et al.(1974), Taitel and Dukler (1976)). Later attempts have focused on finding better correlations for the friction terms where two-phase flow effects have been incorporated into the model. The interfacial shear stress has been mainly modelled by two different approaches for wavy stratified flows. In the most common approach, a global empirical correlation for the interfacial friction factor or the interfacial friction term is obtained from experimental data. Among others, the models proposed by Andreussi and Persen (1987), Andritsos and Hanratty (1987), and Biberg (1999) are based on this method. Analogous to the thin film treatment, the other approach considers the flow above a wavy interface similar to the flow above a rough surface where an empirical correlation for the interfacial roughness is proposed. For waves in deep liquid films, Charnock (1955) linked the surface roughness to the frictional velocity through a non-dimensional parameter. Rosant (1983) modified the Charnock relation for pipe flows. The Charnock relation may be used together with Colebrook's (1939) equation for flows in rough pipes. However, the basic problem still remains - mechanistic models require virtually the same amount of empirical information to obtain the same outcome as purely empirical data correlations even for the simplest flow geometries (Newton and Behnia (2001)). As a consequence, engineers started to explore the possibilities to employ CFD techniques for the prediction of wavy stratified flows. Some investigators have attempted to use the Two-Fluid Model (discussed later), e.g. Akai et al. (1980, 1981), Issa (1988), Liné et al. (1996), and Lorencez et al. (1997), while others have preferred a homogenous approximation of the Two-Fluid Model, in general the Marker and Cell Methods (MAC), for the treatment of free surface problems. At present, the MAC methods can be divided into four, somewhat overlapping categories: Front Tracking, VOF, Level Set and Hybrid methods (McKee et al. (2008)). However, due to a major advantage of mass conservation among others, the VOF and hybrid (mostly based on VOF) methods remain dominant. In the following, the most relevant works are briefly reviewed.
Akai et al. (1980, 1981) applied a two-equation turbulence model to predict the turbulence transport mechanisms in both phase. The effects of the interfacial waves on the flow field were formulated in terms of the boundary conditions for the wavy gas-liquid interface. For the gas phase, the wavy interface was considered to behave as a rough surface in a single phase flow. For the liquid phase, the waves were treated as a turbulence energy source. The numerical results taking into account these effects were in good agreement with the experimental data. However,

the author concluded that the physical considerations were yet by no means perfect and emphasis was placed on more detailed experiments and theoretical approaches.

Shoham and Taitel (1984) proposed a 2D model for the stratified turbulent-turbulent gas-liquid flow in inclined pipes. The interfacial structure is taken into consideration using appropriate correlations for the interfacial shear stress through which the two phases were coupled. The model was capable of predicting the liquid velocity field, hold-up and pressure drop given gas and liquid flow rates, physical properties, pipe size and angle of inclination. Predicted pressure drop and hold-up were tested against available experimental data in the literature, showing substantially better agreement than the predictions of Lockhart and Martinelli (1949) correlation, Taitel and Dukler (1976) and some others.

Issa (1988) presented a method for the calculation of fully-developed, stratified, two-phase flow in inclined channels and pipes. The method includes two versions of the '$k - \varepsilon$' turbulence model, for high and low Reynolds numbers, respectively. Predictions with the technique were compared with the results of other methods and with experimental data for both pipe and rectangular channel geometries where a fairly well agreement was found. The ability of the method to handle inclined flows was also demonstrated. It was concluded that the treatment of the interface played an important role in determining the overall behaviour of the flow and this was reflected in the predicted pressure gradient and hold-up. The merit of the method in including the gas phase in the analysis was highlighted in some of the applications where empirical correlations normally used for the gas phase become invalid. However, the question of what boundary conditions to impose on 'k' and 'ε' at the interface, especially in the presence of waves, remains unresolved. The boundary constraint on 'ε' suggested by Akai et al. (1981) was found to be somewhat unreliable, hence, the introduction of zero-gradient boundary constraint in its place. The author concluded his work by suggesting further examination and assessment.

Lorencez et al. (1997) using the '$k - \varepsilon$' turbulence model, obtained the numerical predictions of his experimental data. Lorencez et al. (1997) pointed out that although the '$k - \varepsilon$' model was originally developed for single phase flows, it has been used to solve some free surface problems by modifying the boundary conditions at the gas-liquid interface (Akai et al. (1981), Celik and Rodi (1984), Issa (1988), Benkirane et al. (1990), Murata et al. (1991) and Lorencez et al. (1991)). A lack of accurate knowledge about the turbulence structure near the gas-liquid interface, however, has prevented formulation of appropriate interfacial boundary conditions, which explains why few studies of free surface turbulent flow problems have been conducted with a two-equation turbulence model. The authors solved this problem by assuming that the convection provided by the secondary flow is negligible. Accordingly, a new interfacial boundary condition for the turbulent kinetic energy for wavy stratified flows was proposed, in which the contributions from the interfacial shear and wave motion to the turbulent kinetic energy were separately considered. Numerical predictions using the interfacial condition produced satisfactory results for both co- and counter-current wavy stratified flows.

Newton and Behnia (2001) used a low Reynolds number '$k - \varepsilon$' turbulence model to extend their previous work (Newton and Behnia (2000)) to account for the interfacial waves. A simple empirical shear stress distribution was imposed on the interface to modify their model for wavy stratified flows. The interfacial roughness of Meknassi et al. (2000) was introduced in the wall functions. Their study extended the work of Liné et al. (1996) to circular pipes and the effect of secondary flows was included using an anisotropic turbulence model.

Mouza et al. (2001) employed a CFD code to obtain the detailed characteristics of horizontal wavy stratified flows for two geometries (pipe and channel). By treating the gas and liquid phases separately, the calculations were performed using the experimental data on pressure drop and liquid film thickness as inputs. The author stated that the essential flow characteristics can be predicted fairly well if realistic estimates of the flow domain and interface roughness are available. However, this reliance on the experimental data is a drawback of this technique. The author also pointed out that the important issue of dynamic interaction between the phases involving wave generation, evolution and propagation is poorly represented by the equivalent roughness methodology and this probably resulted in a weak secondary flow in the gas phase which was very strong in the experiments.

Berthelsen and Ytrehus (2004, 2005) developed a 2D numerical technique to calculate the fully developed stratified flows. The immersed interface method (Berthelsen, 2004a) was used to treat the interfacial boundary condition properly, where the interfaces were represented by level set functions. The wavy interface was treated as a moving rough wall. Turbulent stresses were modelled using a two-layer turbulence model. The simulation results were compared with the experimental data of Espedal (1998). The results showed acceptable agreement with the measurements. Calculated wall and interfacial shear stresses were also found to compare satisfactorily with the experiments. The interfacial roughness, represented by the Charnock parameter, was estimated from Espedal's experiments. It was reported that the results were slightly sensitive to the choice of Charnock parameter. The level of turbulent kinetic energy was found to be slightly overpredicted in the gas phase and was explained as pointed out by other researchers that probably the interfacial boundary value for the turbulent kinetic energy may be too high on the gas side.

Ghorai et al. (2005) employed the Eulerian model coded in a commercial CFD package to calculate the gas-liquid flows in pipes. Turbulence was modelled using the '$k - \varepsilon$' turbulence model. The interface was assumed as a moving wall and to account for the effect of waves, an experimentally determined equivalent roughness was imposed. The turbulent kinetic energy as well as its dissipation rate was also applied. The velocity profile was compared to the experimental results of Strand and Lopez. The agreement between the experiments and simulations was found to be satisfactory. The author concluded that the concept of interfacial roughness can yield good results. However, the problem remains that the interfacial roughness is determined from a certain experimental setup under certain experimental conditions, hence, like empirical and mechanistic approaches, perform poorly when applied to other cases.

Vallée et al. (2007) simulated an air-water stratified flow by using the standard Eulerian model with the free surface option in a commercial CFD code. Turbulence was modelled separately for each phase using the '$k - \omega$' based shear stress transport (SST) turbulence model. A simple grid dependent symmetric damping procedure proposed by Egorov (2004) was used as the interfacial boundary condition on the turbulence quantity 'ω'. The results compared very well in terms of slug formation, velocity, and breaking. The authors concluded that the qualitative agreement between the prediction and experiment is encouraging and showed that CFD can be a useful tool for studying the horizontal two-phase flows.

Wintterle et al. (2007) focussed on the counter-current flow regime. Turbulence was modelled with an extended '$k - \omega$' model with an additional term which improved the turbulence behaviour in the vicinity of the gas-liquid interface. In the first step, a parameter study was performed for the '$k - \omega$' model which indicated a best practice value for the additional two-phase turbulence term. In the second step, a more universal formulation was introduced. Applying this additional two-phase turbulence term yielded numerical predictions which were in good conformity with the experimental results.

De Sampaio et al. (2008) simulated fully developed gas-liquid stratified flows in horizontal pipes. The Reynolds averaged Navier-Stokes equations (RANS) with the '$k - \omega$' turbulence model were solved using the finite element method. A smooth interface is assumed without considering the effects of interfacial waves. The results indicated a fair agreement with the experimental data. However, the author stressed that a better agreement with the experimental data can only be achieved if the behaviour of the modelled turbulent quantities 'k' and 'ω' at the interface can be known in advance with accuracy.

Apart from the grid dependent methods, in the recent years, developments in the area of Smoothed Particles Hydrodynamics 'SPH' (e.g. Dickenson (2009) and Monaghan (2011)) seems very promising in terms of its application to the shear driven thick film flows. At the Institute für Thermische Strömungsmaschinen, some relevant test cases with considerable interaction between the phases have already been successfully simulated (e.g. Höfler et al. (2011)). Vorobyev et al. (2010) has also shown a good agreement while predicting the overall motion of the free surface, wave propagation and reflection from the tank walls in the incident of a collapse of a centralized water column inside a cylindrical tank.

2.2.3 Summary of Film Flow Modelling

In section 2.2.1, it is concluded that the film flow modelling in bearing chambers is still in its early phase and it is required to investigate other engineering fields dealing with the similar flow phenomena. The literature review in this regard suggests that there have been several attempts to (numerically) model the wavy stratified flows time and again over the last three decades. The modelling of interphasic interaction was given the primary importance. It was found that for turbulent shear driven flows, the gas side velocity gradient which is actually responsible for the interphasic interaction can only be reproduced well with the help of additional boundary conditions specified at the interface. Accordingly, similar to the thin film treatment, the most commonly applied boundary condition makes use of a rough wall analogy where the gas-liquid interface is approximated by a moving wall and waves appearing on the interface as roughness. As mentioned in the last section, several authors reported good agreement with the experimental data; however, the limitations especially knowledge on a number of desired input parameters and case dependency of such a boundary condition are also highlighted. Besides case dependency, Mouza et al. (2001) pointed out that the dynamic interaction between the phases involving wave generation, evolution and propagation cannot be modelled using equivalent roughness methodology. Especially in pipe flows, the wave motion on the film surface is considered to be the major source for generating strong secondary flows in the gas phase, hence, cannot be modelled with roughness methodology as well. Therefore, it can be concluded that the technique of equivalent roughness is only suitable for thin films i.e. where the film thickness is negligible as compared to the height of transporting line cross-section and the cases where the secondary motions are either absent or very weak.

Other authors have argued that the incorrect gas side velocity profile is a result of inadequate knowledge on the behaviour of turbulent transport near the wavy gas-liquid interface and placed their focus on the physics of turbulence modelling in the vicinity of the interface. Accordingly, a number of interfacial boundary conditions on the turbulent quantities were proposed. However, due to the limited and sometimes contradictory experimental knowledge gained by different authors on the behaviour of near interface turbulent quantities, a case independency could not be achieved yet. A point to be noted here is that from the literature review it could not be revealed what the source of the incorrect gas side velocity profile is? In addition to that, a clear answer to which methodology is superior and suitable for bearing chamber typical thick film flows also requires a detailed analysis. These informations are necessary for deriving an appropriate and case independent interfacial boundary condition. Thus, it can be said that in the light of discussions made in this section, the necessity to address these questions is clearly evident.

3 Problem Definition and Concretization of Objectives

In the last chapter, a detailed analysis of the state of the art tools and techniques dealing with the bearing chamber typical oil film flows were presented. The advantages and disadvantages of the later were discussed and the need of fundamental experimental investigations and corresponding numerical modelling was identified. In this chapter, first of all, the basic idea behind such an investigation is briefly revised in the problem definition section followed by the scope of the thesis in the framework of the global aim. The thesis objectives are then concretized.

3.1 Problem Definition

A common feature of all future aero engine concepts is the increase in the maximum turbine inlet temperatures and even higher shaft speeds. This demand, however, results in safety and reliability issues (e.g. oil fire/coking) of the aero engines. Therefore, the development in this direction can only be achieved by designing an effective heat management system. The oil system in aero engines is one such heat management system and plays a vital role in cooling the high speed bearings and the separating housings/chambers around them. To enhance the heat transfer limits of this heat management system, exact and detailed knowledge on the governing flow phenomena is required. Glahn (1995) was the first to characterize the two-phase flow phenomena using a (model) bearing chamber test rig and found that a very dynamic oil film attached to the outer chamber wall was present at all operating conditions. Glahn (1995) also measured the heat transfer coefficient around the chamber housing and established that the oil film played a vital role in cooling the chamber walls. The existence and importance of the oil film was also confirmed by all the following investigations and publications of this area (see section 2.1). Therefore, provisions are necessary to investigate the film dynamics in bearing chambers because any system optimization with regard to the most important design parameters e.g. efficient lubrication, reliable heat management and efficient scavenging requires a detailed knowledge on the film dynamics.

3.2 Scope of the Thesis

In section 2.1.1, the detailed analysis of the circumferential film thickness distributions revealed that the bearing chamber typical films, for the most part, can be considered to follow the thin film assumptions and are fairly well understood. In the vicinity of the scavenge off-takes, the film thickness can exceed a couple of millimetres ($\sim 2-5mm$) where the thin film assumptions are invalid. Moreover, the film in the counter-current regime also behaves differently than in the co-current regime and shows a region of high mixing in the form of a recirculation as shown by state 'B' in Fig. 3.1. Hence, the film in the vicinity of the scavenge off-takes is relatively thick, highly dynamic in nature and is posing a major challenge to the engineers dealing with the bearing chamber layouts. In order to predict the film's behaviour, enhanced and more detailed thick film models are necessary which in turn require detailed informations on the film dynamics. Before any progress can be made in this regard, it is necessary to understand the dynamics of thick films in the co- as well as the counter-current regime and therefore, is set as a focus of the work presented in this thesis.

In the past, detailed experimental rigs, e.g. ITS HSBC, were used to understand the film dynamics which proved to be very useful in determining the typical circumferential distributions. However, regarding the individual characteristics of the film not much progress could be made due to several co-occurring phenomena in complex test rigs. As a consequence, till today, no generalized design rules could be formulated. This fact also indicates that in future probably advanced multiphase numerical techniques (CFD) will emerge as the only tool for bearing chamber layouts. Since the numerical techniques dealing with multiphase flows are still in the development phase, well-defined boundary conditions and appropriate validation are required to properly analyse their suitability for thick film simulations. Therefore, test rigs reproducing the film flow around the scavenge off-takes with well-defined boundary conditions are necessary to deliver the fundamental understanding of the film dynamics as well as appropriate test cases for the corresponding CFD simulations. In chapter 2 (see Fig. 2.19), the basic idea behind a simple test rig is also presented where it is shown that the major forces acting on an infinitesimal film element can be reproduced with the help of a linear rotatable test rig. As will be discussed in the next chapter, the test rig developed in this manner will take the form of a 'Horizontal/Inclinable channel' as shown in Fig. 3.1.

Now recalling the core aim i.e. to be able to simulate complete bearing chambers, the methodology adopted in this thesis (dashed rectangle in Fig. 3.1) lays the fundamental building block in achieving the overall aim. The state 'A' represents the possible flow phenomena in the bearing chambers as understood today followed by the state 'B' which represents the first system abstraction i.e. neglecting the droplet-film interaction in order to be able to investigate the individual features of the film flow. Practically, this can be achieved by keeping the shearing gas flow under the droplet entrainment limit. To fulfil the requirements of well-defined boundary conditions and the definition of a suitable validation parameter, a simple test rig is designed in an iterative CFD procedure where all the requirements can be sufficiently met and at the same time ensuring near (scavenge) off-take similar film flows. The data acquired, will be used to setup a reliable CFD model and validate the simulation results. On successful realization of the CFD model, the methodology will be ready to simulate the oil film phenomena in the bearing chambers as shown by the state 'C' in Fig. 3.1. In the next (future) step, the test rig can be extended to investigate droplet stripping and droplet-film interactions, hence, providing the necessary data to setup and validate a CFD model with droplets in the system. The state 'D' represents a full bearing chamber simulation which will be the result of the CFD methodology developed in the present and future works.

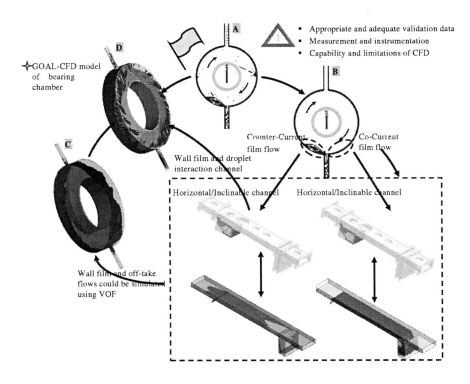

Fig. 3.1: Graphical depiction of thesis methodology and relation to the overall objective

3.3 Concretization of Objectives

In the light of scope of the thesis, the thesis objectives can be divided and concretized into an experimental and a numerical part. The work which is carried out in the experimental part of this thesis is as follows:

- Design and construction of a test rig which ensures similarity to the thick films present in the vicinity of the scavenge off-takes

- Provisions for co- & counter-current investigations in the test rig

- Effective isolation and superposition of driving forces (shear and gravity)

- Well-defined and reproducible boundary conditions

- Definition of a new parameter to effectively quantify the thick film dynamics and to properly analyse/validate the multiphase models

- Good optical access to allow for the qualitative analysis of thick film dynamics

- Investigation of the oil film behaviour under engine relevant conditions i.e. high temperatures and pressures

- Variation of parameters
  - film loading,
  - interfacial shear stress &
  - gravitational influence

The work which is carried out in the numerical part of this thesis is as follows:

- Analyse the state of the art multiphase models in the light of suitability and feasibility

- Selection and/or development of a suitable multiphase model to provide a CFD tool applicable to thick films

- Validation of the CFD methodology against the test cases made available through the experimental investigations in this thesis

- Application and validation of the CFD methodology with a more engine relevant test case - ITS model bearing chamber

4 Experimental Investigations of Thick Film Flows

In the previous chapters, the need to investigate the thick film flows in an environment isolated from the co-occurring flow phenomena in bearing chambers is clearly identified. Pursuing this aim, it is shown in Fig. 2.19 that a linear test rig can be used to approximate the film flow around the scavenge off-takes. Accordingly, in this chapter a carefully designed linear test rig is presented which is used to carry out the necessary fundamental investigations. The special design of the test rig allows for a sufficient consideration given to all the prerequisites and demands made on the test rig in section 3.3. The design features of the test rig are discussed in the following sections. Later in the chapter, the experimental results are presented and analysed. The results are then discussed in relation to a possible characterization of the thick film flow in bearing chambers.

4.1 Stratified Flow Test Rig

To achieve similar films, the forces acting on the film, film thickness and operating conditions are set as the similarity criteria between the ITS HSBC test rig and the stratified flow test rig as shown in table 4.1. The ITS HSBC test rig can rotate up to $16,000rpm$. However, from

Similarity Criteria	Unit	ITS HSBC Test Rig	Stratified Flow Test Rig
Core gas velocity[1]	m/s	$4(6) - 16(24)$[2]	$4 - 12$
Gravitational influence	°	φ	see Eq. 2.5
Film thickness	mm	$2 - 5$	$3 - 6$[3]
Temperature	$°C$	100	[4]
Pressure (relative)	bar	$0 - 3$	[5]

[1] simulates similar interfacial shear stress
[2] core gas velocity is $20(30)\%$ of the shaft ($\varnothing 128mm$) tangential velocitys
[3] without shearing air flow
[4] achieved by matching the film physical properties
[5] discussed later

Table 4.1: Similarity criteria between the ITS HSBC and the stratified flow test rig

$12,000rpm$ onwards, the counter-current flow regime vanishes and the test rig demonstrates film thicknesses less than a millimetre (see section 2.1.1) which can be regarded as a fairly thin film. Therefore, in table 4.1, the core gas velocity is calculated for the range $3,000 - 12,000rpm$ using the approximation $(20 - 30\%)$ known from literature (e.g. Gorse et al. (2003), Morrison et al. (1991)). Hence, in the stratified flow test rig, similar interface shear stress is achieved by blowing the bearing chamber relevant air velocities over the liquid film with a minimum possible initial film momentum for the relevant film thicknesses. The upper limit on the superficial gas velocity '$u_{SG} \sim 12m/s$' is restricted by the onset of liquid entrainment in the gas phase (droplet shedding).

The rig has a rotational degree of freedom of up to $45°$ to analyse the cases under pure gravitational effects as well as the superposition of gravitational and shear effects. In this manner, the test rig allows for an effective isolation of the driving forces; hence, the influence of an isolated force on the film dynamics can be investigated. The similar film thickness and gravitational influence will ensure similarity in the film surface structure and gravitational force. This implies that the air will experience a similar surface roughness as in the bearing chamber environment.

A combination of the design of a pre-filming device (discussed later), test rig dimensions and carefully selected film loadings, keep the film thicknesses similar to the near scavenge off-take film thicknesses $(2 - 5mm)$. In table 4.2, the required film loadings for the relevant film thicknesses are presented.

Film loading *'V̇'* $[l/min]$	*Film thickness*[1] *'δ'* $[mm]$
5.07	~ 3
6.22	~ 4
8.34	~ 5
9.80	~ 6

Table 4.2: Required film loadings for the relevant film thicknesses

To investigate the behaviour of an oil film at high temperatures, the physical properties of the film are modified to match the oil properties during operation. In order to understand the role of an individual physical property on the behaviour of film dynamics, physical properties are also modified and investigated independently in the experiments.

In this work, glycerine is selected to match the viscosity of water to that of engine oil at $100°C$. Glycerine is used because it is of almost no toxicity and has a surface tension $(0.064N/m)$ very close to that of water $(0.072N/m)$. In addition to that, glycerine has a very useful property that when added in an appropriate proportion to water it increases the viscosity of water in a wide range with only a slight change in the surface tension (see Fig. 4.1). In the past, many researchers have also successfully employed glycerine for this purpose.

In practice, some surface active agents e.g. Tween 80, butanol etc... are used to alter the surface tension of water without affecting its viscosity. However, it is very difficult to achieve a surface tension as low as $0.023N/m$ (or below) which the oil has at $100°C$ $(200°C)$. The surface active agents, when mixed with water to reduce its surface tension, also produce foam as a by-product. This is an inherent characteristic of almost all surface active agents and a major disadvantage in the type of investigations planned in this research work. Therefore, after testing a couple of surface active agents, a common vinegar based cleaner is used to reduce the water surface tension by approximately 50% and at the same time producing negligible foam.

---

[1]Film thickness is determined at the start of the test-section without a co-current air flow and by using an immersed scaled pin

Glycerine and vinegar based cleaner are added simultaneously to water to match both the proper-

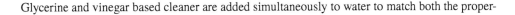

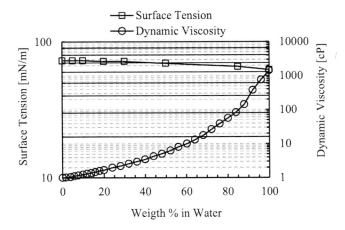

Fig. 4.1: Properties of aqueous solution of glycerine (Kihm (1996))

ties to that of engine oil during operation (see Table 4.3).

Temperature [°C]	Aero Engine Oil		Water[2] + Glycerine[3] + Surfactant (at room temperature)	
	$\sigma\,[N/m]$	$\nu[cSt]$	$\sigma\,[N/m]$	$\nu[cSt]$
100	0.023	5.2	0.0365	5
200	0.015	1.4	0.0365	1.2

Table 4.3: Properties of oil and investigated fluid at engine relevant conditions

The influence of pressure on the film dynamics can be investigated by analysing the means by which the interphasic interactions take place. A force balance on a fully developed film flow reveals that the interfacial shear stress, given by the relationship '$\tau_i = 0.5 f_i \rho_G (u_{SG} - u_{SL})^2$', is the only means of interaction between the phases where pressure does not appear explicitly rather its influence is indirectly included in the gas density. However, care should be taken when transferring the results to high pressure systems because it can be shown mathematically that for an increase in density by a factor 'α' i.e. '$\rho = \alpha \rho_G$', the upper limit (droplet shedding) on the superficial air velocity 'u_{SG}' decreases approximately by '$\frac{1}{\sqrt{\alpha}}$'. The interfacial shear stress at atmospheric and elevated pressures can now be written as follows

$$\tau_i = 0.5 f_i \rho_G (u_{SG} - u_{SL})^2 \ at\ atmospheric\ pressures \qquad (4.1)$$

$$\tau_{i,\alpha} = 0.5 f_i \alpha \rho_G (u_{SG} - u_{SL})^2 \ at\ elevated\ pressures \qquad (4.2)$$

---

[2]The viscosity of oil at $200°C$ is comparable with pure water at room temperature

[3]Glycerine is only required to match the properties of oil at $100°C$

Based on the discussions in section 2.1.1, it can be stated that the liquid superficial velocity 'u_{SL}' is negligible as compared to that of gas. Neglecting the superficial liquid velocity and comparing Eq. 4.1 & Eq. 4.2 gives an estimate of the decrease in superficial air velocity for the same shear stress exerted by the gas on the gas-liquid interface at elevated pressures. Therefore, in the fundamental investigations presented here, it is assumed that the influence of pressure does not have to be investigated separately.

$$\Rightarrow u_{SG}^2 = \alpha u_{SG,\alpha}^2 \tag{4.3}$$

$$\Rightarrow u_{SG,\alpha} = \frac{u_{SG}}{\sqrt{\alpha}} \tag{4.4}$$

4.1.1 Test Rig Design and Boundary Conditions

The design focus of the new test rig is to avoid the geometric similarity to the ITS HSBC test rig, yet, allowing insights into the dynamics of thick oil films and provide the necessary boundary conditions and suitable validation data for the numerical simulations. For this purpose, the ITS stratified flow test rig constitutes of a transparent modular construction made of Plexiglas with a rectangular cross-section ($192mm \times 50mm$) and a test section length of approximately $915mm$ (see Fig. 4.2). The small length of the test section ensures undeveloped flow conditions, yet, it is long enough to investigate the behaviour of the film flow. The large aspect ratio (width to the channel's height) of the test rig minimizes the wall effects on the film flow. The rig is primarily designed to deliver pressure drop and film thickness measurements along the channel length. The top walls of the rig modules (inlet '1', test section '3' & outlet '5') are provided with circular openings of diameter $\varnothing 100mm$ each, where additional instrumentations can be installed. Due to the rig's excellent optical access, the gas-liquid interface characteristics can easily be investigated qualitatively and the investigations can be extended in future to analyse the local flow field quantitatively (e.g. with the aid of some suitable optical techniques). The rig's modular construction also allows for a variable exit boundary condition e.g. single/separate outlet for phases as well as an alternate (e.g. larger) size of the liquid separator.

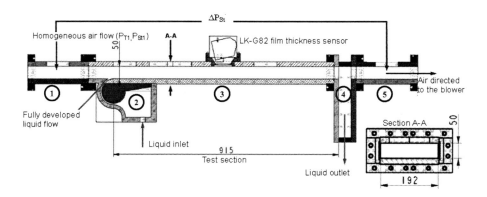

Fig. 4.2: ITS stratified flow test rig

The required air flow in the test rig is provided by a blower of type RD 5 ($Elektror^{®}$, $max.$ $10m^3/min \sim 17m/s$) which sucks the surrounding air through a bell-shaped intake into the inlet module '1' and then releases it again to the surrounding (see Figs. 4.4 a-b). The junction between the bell-shaped intake and the inlet module '1' is further provided with a fine grid ($1mm \times 1mm$) to assist an intense mixing of the incoming air flow. To control the flow of air through the test section, a throttle valve on the suction side of the blower is used. In the middle of the inlet module '1', the total pressure of the air flow is measured with the help of a vertically traversable Pitot tube and the static pressure with the help of a surface levelled fine hole. Two differential manometers of type MP 200 M ($KIMO^{®}$, $range \pm 2500Pa - Tol. \pm 2Pa$) are used for this purpose. Another differential static pressure is taken between the inlet and outlet modules using the type MP 200 P ($KIMO^{®}$, $range \pm 500Pa - Tol. \pm 0.8Pa$). The purpose of this differential static pressure measurement will be explained in the next section. Taking into account the fact that the blower provides a fluctuating air flow, each pressure measurement is averaged over a minimum time interval of $30secs$.

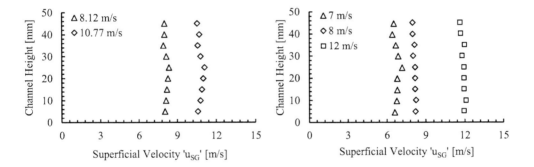

Fig. 4.3: Block profile inlet air flow. Co-Current (Left); Counter-Current (Right)

The superficial air velocity distribution in the inlet module can be acquired from the corresponding dynamic pressure distribution along the channel height by traversing the Pitot tube. Considering the fact that suitable measures are taken to ensure intense mixing, a block profile superficial air velocity is achieved at the inlet as can be seen in Fig. 4.3 (Left). By interchanging the bell-shaped intake of the inlet module with the rectangular to circular section adapter of the outlet module, the modular construction of the test rig also allows for the realization of counter-current configuration using the same test rig and components (see Fig. 4.4b). Using the circular openings provided in the top walls of all the modules, the position of the instrumentation can also be adapted accordingly (see Figs. 4.4 a-b). The superficial air velocity distribution in the counter-current configuration of the test rig also indicates an almost homogenously mixed block profile (Fig. 4.3 (Right)). In this way, a well-defined boundary condition is achieved for the air flow in the co- as well as the counter-current configuration of the test rig.

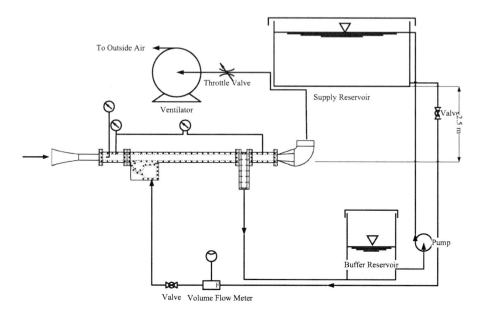

Fig. 4.4a: Co-Current flow loop

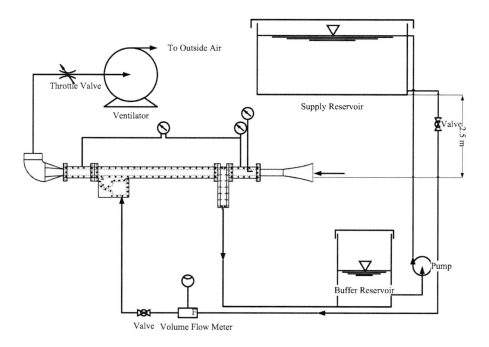

Fig. 4.4b: Counter-Current flow loop

A 200 litre liquid reservoir, placed approx. $2.5m$ above the test rig, supplies the required volume flow rate. The supply pipe is provided with a G2 type ($GPI^{®}$, $range\ 3.8 - 37.9l/min - Tol. \pm$ 1.5%) turbine volume flow meter. A digital computer attached to the flow meter displays the instantaneous flow rate of the passing liquid. The liquid flow rate can be set to a desired value by adjusting a lever valve installed after the volume flow meter. Liquid enters the test section with the help of a compact pre-filming device (see Fig. 4.5 (Left) & Fig. 4.2 '2') designed in an iterative CFD procedure. A manifold liquid inlet connected to the buffer volume of the pre-filming device significantly reduces the momentum of the entering liquid jet and results in almost a stagnant liquid reservoir (plenum) in the pre-filming device. The V-shaped top wall of the reservoir has an increasing slope which drives the trapped air and bubbles to deposit on the highest point where it can be vented with the help of an opening. A cylindrical film channel connected to the reservoir intersects with the test section '2' (see Fig. 4.1) tangentially to allow for an uninterrupted gas and liquid flow in the test section. The film channel is constructed with the help of two coaxial cylinders with diameters, $\varnothing 100mm$ & $\varnothing 110mm$ respectively, selected in a manner to give a constant channel width of $5mm$. The film channel is higher than the reservoir and therefore, allows for filling the reservoir without spilling the liquid in the test section. In this way, it can be ensured that only liquid is present in the reservoir before the experiments can begin. After the reservoir is filled, pumping the liquid results in a uniform and homogenous liquid film entering the cylindrical film channel as is evident from the streamlines and velocity distribution shown in Fig 4.5 (Right). Thus, a very well-defined and easy to implement inlet boundary condition is achieved for the liquid film entering the test section. After passing through the test section, the liquid is collected in the separator tank '4' (see Fig. 4.2) which is always filled with some liquid. The collector tank is continuously drained out at the same rate as the liquid enters the test section. This helps in achieving a constant liquid level in the tank and also keeps the complete air flow directed towards the outlet module with minimal disturbances. In addition, this technique also provides a well-defined outlet boundary condition for the film flow leaving the test section. Due to the separation of phases, only a single phase (air) is present in the outlet module '5' (see Fig. 4.2) where a static pressure measurement can easily be carried out, hence, providing a well-defined outlet boundary condition for the air flow leaving the outlet module.

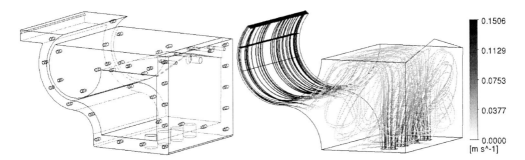

Fig. 4.5: Pre-filming device (Left); CFD streamlines & three surfaces in the film channel coloured with flow velocity (Right)

The drained liquid from the separator tank '4' accumulates in a buffer reservoir (see Fig. 4.4a-b) so long until the recirculating pump is activated by a signal transmitted by an electronic floating valve mounted in the supply reservoir. This usually happens even before the liquid level can experience a decrease of $20mm$. Therefore, it can be ensured that the liquid level in the supply reservoir remains sufficiently constant. Together with the pre-filming device, this allows for an uninterrupted constant volume flow rate, hence, constant initial film conditions in the test rig.

4.1.2   Definition of a New Validation Parameter

A major difference between the investigations presented in this thesis to most of the prior published studies is that the film dynamics are not studied directly but by considering how the air experiences the film. If the film was analysed directly, only local quantities like film thickness/velocity profile could be measured at a point or few which are neither sufficient for describing the global behaviour nor can be used to fully characterize the undeveloped thick film flows. Moreover, the multiphase numerical techniques are still in the development phase and require an extensive global validation before a detailed validation of the local quantities can take place. Therefore, for effectively quantifying the influence of the driving forces, a new integral approach is introduced which accounts for the complete momentum transfer between the liquid film and the air flow. This approach is also very effective in the sense that it gives a single and easy to measure parameter characterising the complete undeveloped wavy stratified film flows. The concept behind the measurement of the complete momentum transfer from the fast moving air flow to the low velocity liquid film is based on the fact that the test rig is designed in a manner that air interacts with the liquid film only in the test section. In the inlet and outlet modules, where the simultaneous pressure measurements are carried out, the liquid film is not present. Therefore, the total pressure drop 'ΔP' in the test section can be given - with sufficient accuracy - by the differential static pressure 'ΔP_{St}' because the dynamic pressure (integrated over the cross-section) is constant in the inlet and outlet sections (same cross-section and air volume flow rate). Therefore, it follows:

$$\Delta P \approx \Delta P_{St} \qquad (4.5)$$

The pressure drop measurements are repeated with and without a film in the test section. The pressure drop 'ΔP_{WF}' with a film in the test section is always greater than the pressure drop without a film 'ΔP_{WoF}' in the test section. The pressure drop recorded in this manner, in addition to the momentum transferred to the film, also constitutes of other components as shown in Fig. 4.6; however, considering the fact that air has a very low viscosity ($\sim 1.8 \times 10^{-5} Pas$), it is safe to assume that almost the entire pressure drop goes in the momentum transfer. Therefore, the additional pressure drop 'I' caused by the film in the test section, as compared to the channel walls, gives the complete momentum transferred to the film from the shearing air flow and can be expressed by Eq. 4.6.

$$I = \Delta P_{WF} - \Delta P_{WoF} \qquad (4.6)$$

In contrast to the local film thickness measurements which can only be performed at limited locations, the momentum transfer is a global parameter yet very easy to determine. A comprehensive quantitative analysis can now be performed efficiently and effectively using this parameter.

Another benefit is, since only air is measured to determine the momentum transferred to the film, the data presented here depends only on the physical properties of air and is independent of the film physical properties. As will be shown later, due to the possibility of different film surface structures (waves) arising from the different film properties, an offset may occur; nevertheless, the trend of the momentum transfer will remain. It can now be concluded that all the prerequisites and demands made on the test rig in section 3.3 can be fulfilled to a satisfactory level and the desired fundamental investigations can now be performed using this test rig.

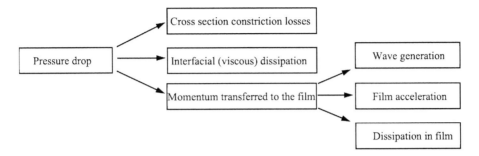

Fig. 4.6: Distribution of measured pressure drop

4.2 Experimental Results and Discussion

4.2.1 Experimental Procedure

The experiments are performed by setting a defined air flow rate over a defined liquid film loading. The momentum transferred to the film is calculated in the same manner as described by Eq. 4.6. The air flow rate is then increased in steps and it is noticed that a reproducible trend exists in the momentum transfer curves. The existence of a trend is explained on the basis of losses in a single phase turbulent air flow which is later shown to be physically plausible.
A qualitative analysis of the film surface is also performed where characteristic waveforms for the investigated regimes (co- & counter-current) are found and discussed in relation to the quantitative analysis and literature. The quantitative and qualitative analysis of water is presented first which serves as the reference case. The influence of physical properties is discussed in relation to the reference case using only the minimum and the maximum film loadings for the reasons stated later. Only one inclination angle (usually the max. possible) is used to explain the supplementary action of gravitational forces.

4.2.2 Reproducibility Study

Reproducibility is ensured by repeating every data point (see point scatter in Fig. 4.7), first, by keeping the liquid flow rate constant and setting a gas flow rate which is then reached by decreasing higher gas flow rates and by increasing lower gas flow rates. Second, the gas flow rate is kept constant and the liquid loading is reached from higher as well as lower loadings. The reproducibility study can be summarized as follows:

- Varying gas flow rate and keeping the liquid flow rate constant leads to reproducibility within a small tolerance of $\pm 1Pa$. In the typical measured range of $10 - 130Pa$, this corresponds to an error of approx. $10 - 0.8\%$

- Varying liquid flow rate and keeping the air flow constant leads to high reproducibility

A good reproducibility indicates that the introduced technique is generally valid for analysing undeveloped multiphase flows of the discussed nature.
Note: Point scatter will only be shown once for the reference case. Later, only polynomial representation of the data will be provided.

4.2.3   Co-Current Flow Regime

In this flow regime, the driving forces are acting in the same direction (see Fig. 2.18). The horizontal co-current flow regime is of little importance for bearing chambers; however, it can be used to effectively appreciate the influence of gravitational forces on the thick oil films.
The pressure drops measured in the gas flow without and with a liquid loading $(5.07 - 9.8l/min)$ are plotted against the superficial gas velocity in Fig. 4.7. The curves are a 2nd degree polynomial fit $(R^2 > 0.99)$ of the measured data points. The choice of a second order polynomial is based on the physical reason that the pressure loss in a turbulent flow is proportional to the square of the mean flow velocity. This behaviour remains true in the presence of a liquid film in the test section suggesting that the air flow experiences the liquid film as an additional resistance leading to an additional pressure loss. Accordingly, with increasing film loadings, the resistance in the test section also increases as can be seen in Fig. 4.7 (summary) where the pressure drop curves shift towards higher pressure losses. This fact is also in agreement with the observation made by Flouros (2005) who reported an increase in the power consumption with an increase in the oil flow rate in a bearing chamber test rig (see Fig. 2.15). To understand the behaviour of increasing pressure drops in air with an increase in the film loading, a step by step analysis of the underlying physical phenomena can be used. From the aspect of air flow in the test section it can be said that when a film enters the test section, it has a finite thickness and the air sees a reduction in the channel cross-sectional area. If the film loading is now increased and it is assumed that the film thickness remains constant then the film velocity must increase leading to a decreased velocity difference between the phases. If this is true then the pressure drop should be lower for higher film loadings than for lower film loadings. Since this is not the case here, it can be stated that an increase in the film loading also causes an increase in the film thickness; consequently, an increase in the film inertia to undergo acceleration and interface deformation. Therefore, the resistance in the test section also increases causing a corresponding increase in the pressure drop. This hypothesis can also be confirmed from the observations made in the ITS HSBC test rig where it was shown that an increase in the film loading was always coupled with an increase in

the film thickness (see e.g. Figs. 2.4, 2.9 & 2.12).

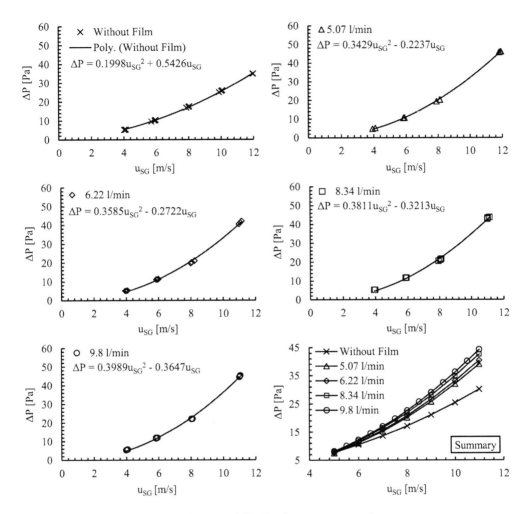

Fig. 4.7: Influence of film loading on pressure drops

In Fig. 4.8, a number of pictures with increasing gas velocities are provided to assist a qualitative analysis of the investigated film loadings. The pictures captured the front view of the test section with the camera slightly facing down to cover the entire width of the test section. The behaviour of the gas-liquid interface follows the concept of a critical lower limit on the gas velocity given by Hanratty et al. (1957) below which the film surface remains flat. The first transition from the flat/smooth gas-liquid interface to the inception of very small surface instabilities happens at gas velocities around $4\,m/s$. However, these small surface instabilities are only recognizable for the two higher film loadings and cannot be categorized as the regular 2D waves as reported in several former studies (e.g. Hanratty et al. (1957), Cohen et al. (1965)).

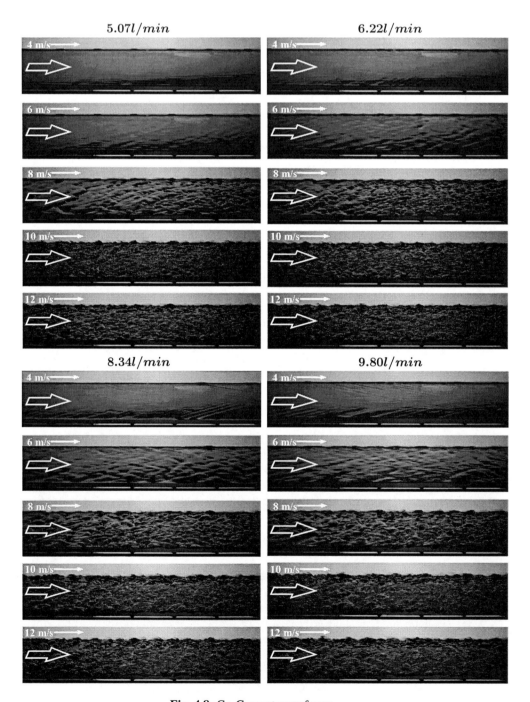

Fig. 4.8: Co-Current waveforms

From approximately $6m/s$ onwards, pebbles start to form on the film surface and the waves can be regarded as 3D in nature. This is also in agreement with the observation made by Hanratty et al. (1957). Cohen et al. (1965) also regarded these waveforms as 3D and defined them as the waves with equal characteristic lengths in the direction parallel and perpendicular to the flow. As the gas velocity is increased, it is observed that a distorted 2D wave occupying the entire channel width appeared (see 3rd picture from top at $8m/s$ especially for the lowest film loading in Fig. 4.8) which after a certain entrance length divides and multiplies into a number of squalls/pebbles. This happens around $7 - 8m/s$ gas velocity. With increasing shear on the film surface, the amplitude and frequency of the waves increases with a decrease in the entrance length. Further increase in the gas velocity at times resulted in waves (roll waves) rolling over the film and droplets tearing-off from their peaks. This point usually occurred around a gas velocity of $12m/s$ and therefore, is not dealt within the scope of this thesis.

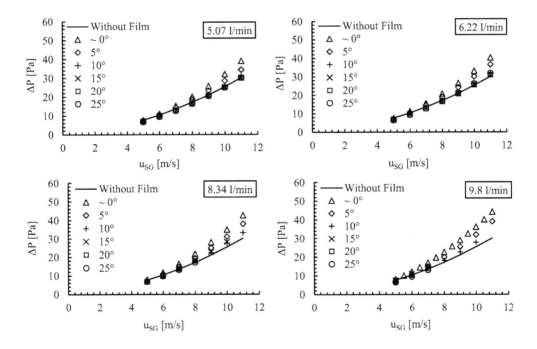

Fig. 4.9: Influence of inclination on pressure drops

Fig. 4.8 also shows that increasing the film loading has no considerable effect on the waveforms generating and propagating on the film surface. The difference lies only in the wave amplitude which appears to be higher, due to a higher momentum transfer (see Fig. 4.7 (summary)), for a higher film loading at the same gas velocity. As mentioned earlier, to simulate the bearing chamber typical wall film flows gravitational effects must be included. Therefore, the investigations carried out in the horizontal test rig configuration are complemented with tests inclining the rig up to $25°$ in intervals of $5°$ (see Fig. 4.9). From $15°$ onwards no considerable difference can be recorded in the measured parameter (momentum transfer), therefore, inclination angles larger than $25°$ are not investigated.

The plots indicate that for thick films, even a small inclination angle of $5°$ has a noticeable effect on the pressure drop (see $\sim 0°$ in Fig. 4.9). With increasing inclination angles the pressure drop continuously decreases. From $15°$ onwards, no further change in the pressure drop can be recorded and the distributions almost overlap. This behaviour remains even with increasing gas velocities. The lack of data for higher inclination angles and gas velocities is due to the fact that the liquid film starts to jump (due to gravity induced additional momentum) over the separator tank, hence, further measurements can not be taken. The behaviour of the film however remains consistent, therefore, the curves can be extrapolated as long as the film remains sub-critical (no droplet shedding).

For a qualitative analysis pictures corresponding to the investigated film loadings are presented in Fig. 4.10. The film surface deforms only slightly in the inclined case; therefore, only pictures corresponding to the highest investigated superficial gas velocity ($\sim 12m/s$) are shown to demonstrate the strong influence of gravitational force on thick films. It can be observed qualitatively that increasing the inclination angles result in a decrease in the wave amplitude and frequency i.e. an increase in the wave length. When compared to the corresponding horizontal configuration where $12m/s$ can be seen as the droplet shedding limit, the film remains very stable with no sign of droplet shedding. Hanratty et al. (1957) and Cohen et al. (1965) also reported that thinner films are more stable than thick films both in terms of the surface instabilities and droplet shedding limit. As the inclination angle increases, the entrance length for the generation of waves increases because the film acceleration results in a decreased interfacial shear stress. The wave generation mechanism, however, remains consistent with the horizontal case i.e. a single 2D wave which spreads along the width, distorts and divides into a number of smaller waves (see 2nd picture from top e.g. for $5.07 l/min$ in Fig. 4.10). For all film loadings, it can be qualitatively confirmed that from $15°$ onwards no significant change takes place on the film surface. The quantitative analysis also revealed in Fig. 4.9 that from $15°$ onwards the pressure drop curves start to overlap. The pressure drop depends on the velocity difference between the phases and the surface roughness; therefore, it can be said that these two characteristics of the test section remains constant. In other words, this implies that the film velocity (hence, the film thickness) and surface roughness in the test section remains constant. This situation where the film parameters (velocity and surface roughness) do not change anymore is referred to as 'constant film dynamics' in this thesis.

Other characteristics of the film flow can also be determined by first analysing the isolated influence of the gravitational force, i.e. without a co-current air flow, on the film dynamics. Accordingly, it can be said that a continuous increase in the inclination angle imparts a continuous gravitational-induced momentum to the film. For a constant film loading, the film experiences a continuous decrease in the thickness with a corresponding increase in the velocity. At a critical inclination angle of about $15°$, the film thickness decreases to a point where the viscous forces become equal to the gravitational forces i.e. a transition from thick to thin film takes place. Further increase in the inclination angle has no influence on the film thickness and the film can be regarded as both laminar and fully developed in the test section.

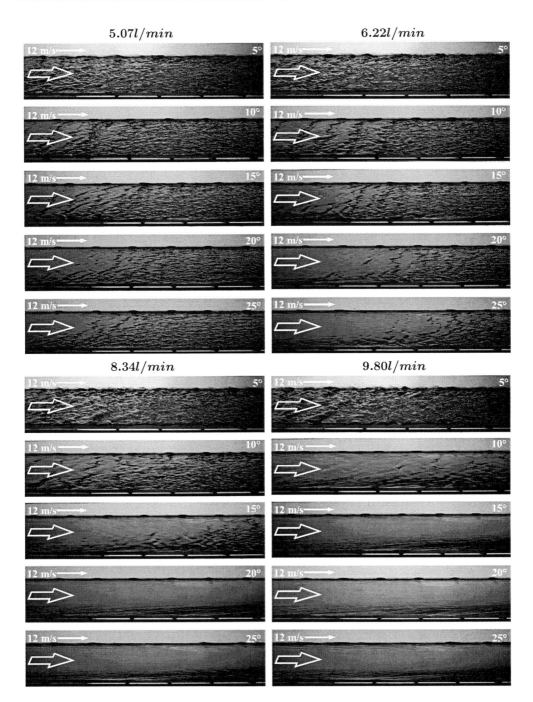

Fig. 4.10: Co-Current waveforms under gravity

In table 4.4, the average film thickness data, without a co-current air flow, with increasing inclination angles is reported. The film thickness is determined at three points (start, middle and end) along the channel length with the help of an immersed scaled pin inserted from the circular openings in the top wall of the test section. It can easily be determined from the film thickness data that the gravitational forces drive the film very quickly to the supercritical flow regime ($Fr > 1$). Now returning to the original problem of shear driven films, it can be stated that the flow phenomena explained above are only enhanced by the fast moving gas flow; hence, the film remains laminar, fully developed and supercritical in nature.

Angle	5.07l/min	6.22l/min	8.34l/min	9.80l/min
5°	$2-3mm$	$3mm$	$3-4mm$	$4mm$
10°	$2-2.5mm$	$2.5-3mm$	$3-4mm$	$4mm$
15°	$1-1.5mm$	$1.5mm$	$1.5-2mm$	$2mm$
20°	$1-1.5mm$	$1.5mm$	$1.5-2mm$	$2mm$
25°	$1-1.5mm$	$1.5mm$	$1.5-2mm$	$2mm$

Table 4.4: Average[4]film thickness distribution with increasing inclination

4.2.3.1 Influence of Surface Tension

To demonstrate the influence of reduced surface tension on the film dynamics, the pressure drop incurred is compared to the pressure drop incurred by the reference case (water - high surface tension). Since the pressure drop curves exhibit similar trends (also 2nd order polynomial), therefore for a clarity purpose, only the curves corresponding to the minimum and maximum film loadings i.e. $5.07l/min$ and $9.8l/min$ are presented. The supplementary influence of the gravitational force is also accounted for by repeating the same experiments at an inclination angle of $15°$ (see Fig. 4.11, Right). The reason why this cannot be done for the highest investigated inclination angle ($25°$), as is the case for all later investigations, is as follows. Keeping in mind the distribution of momentum transfer provided in Fig. 4.6, it can be said that due to a lack of waves on the film surface, the film is slightly faster than in the cases with interfacial waves; hence, the liquid film starts to jump relatively earlier over the separator tank. This is also the reason why the air velocity is restricted to $9m/s$ for the investigations with the maximum film loading (see Fig. 4.12b, right).

---

[4]Average of the film thicknesses measured at three points (start, middle and end)

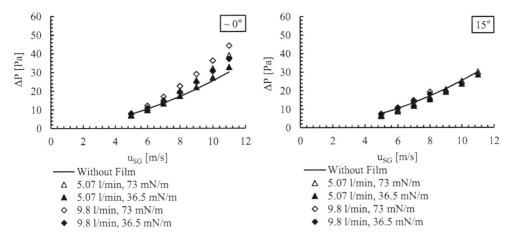

Fig. 4.11: Influence of low surface tension on pressure drops

The plot in Fig. 4.11 (left) indicates that in the horizontal case the pressure drop decreases with decreasing surface tension for all film loadings. However, the general trend that a higher film loading incurs a higher pressure loss, as in the reference case (water), remains the same.

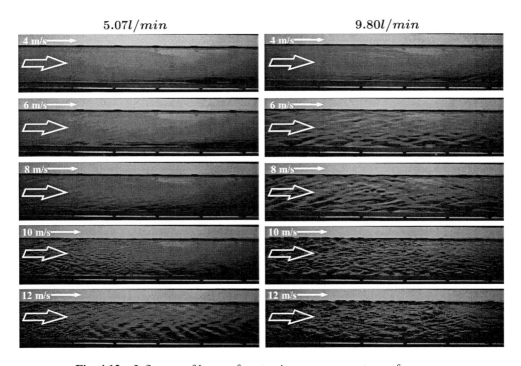

Fig. 4.12a: Influence of low surface tension on co-current waveforms

From the corresponding pictures provided in Fig. 4.12a, it can be seen that reducing the surface tension results in relatively small surface instabilities (waves) as compared to water (Fig. 4.8). This observation is in accordance with some other authors as well e.g. Hanratty and Engen (1961), Weisman et al. (1979) etc... who have reported a similar trend but did not deliver any explanation. The small pressure drop in this case can also be linked to the relatively flat surface. However, to understand the underlying physical phenomena, a detailed investigation of the local flow field is necessary (and is recommended) before a statement can be made in this regard.

The effect of gravitational forces can be seen quantitatively in the right plot of Fig. 4.11 and qualitatively in Fig. 4.12b. It can be said that in the investigated range, surface tension plays no considerable role in gravity driven films i.e. the pressure drop is negligible and matches that incurred in the reference case. This is true because the increasing influence of gravity accelerates the film resulting in a thinner film until viscous forces become dominant. The acceleration of the film also results in a smaller relative velocity between the phases. These factors result in similar film dynamics (film's velocity/thickness and surface roughness) in the cases with low as well as high surface tensions (reference case), hence, similar pressure drops.

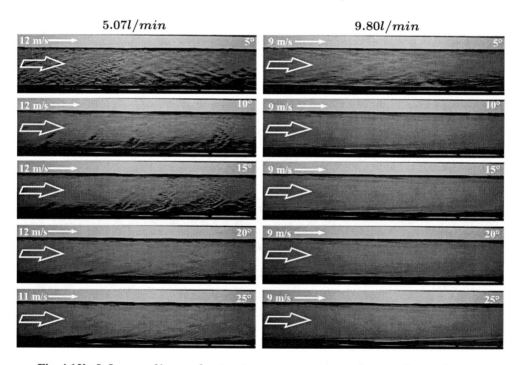

Fig. 4.12b: Influence of low surface tension on co-current waveforms under gravity

4.2.3.2 Influence of Viscosity

The plots indicate that increasing the viscosity results in a somewhat increased pressure drop as compared to the reference case. This can be attributed to the relatively thicker film which results when the film, under the action of relatively high viscosity (almost 5 times the reference case), slows down in the test section. As a result, the relative velocity between the phases increases increasing the interfacial shear which is compensated by bigger waves on the film surface.

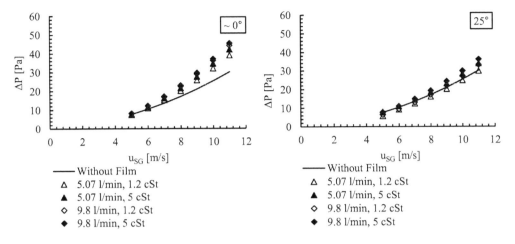

—— Without Film
△ 5.07 l/min, 1.2 cSt
▲ 5.07 l/min, 5 cSt
◇ 9.8 l/min, 1.2 cSt
◆ 9.8 l/min, 5 cSt

—— Without Film
△ 5.07 l/min, 1.2 cSt
▲ 5.07 l/min, 5 cSt
◇ 9.8 l/min, 1.2 cSt
◆ 9.8 l/min, 5 cSt

Fig. 4.13: Influence of high viscosity on pressure drops

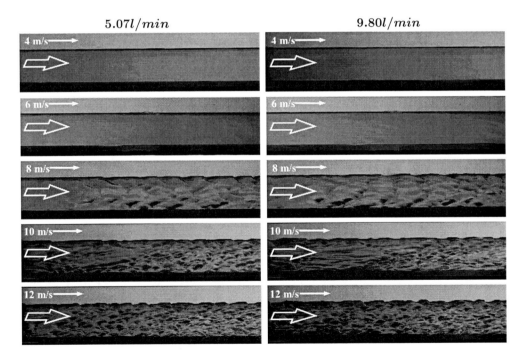

Fig. 4.14a: Influence of high viscosity on co-current waveforms

In accordance with the reference case, the pressure drop decreases with increasing inclination angles; yet, it remains higher for the high viscous liquid unlike the case with low surface tension where they become similar. The viscosity has a dampening effect on the wavy interface (see section 2.1.2.1 for details). This is evident from the second picture of Fig. 4.14a where almost a flat surface can be seen as compared to the reference case. When the waves appear, the entrance length is usually longer than for the reference case. Besides these differences, other difference can neither be observed in the surface waves nor concerning their generation mechanism. The same is true for the visual observations, Fig. 4.14b, made during an increase in the inclination angles.

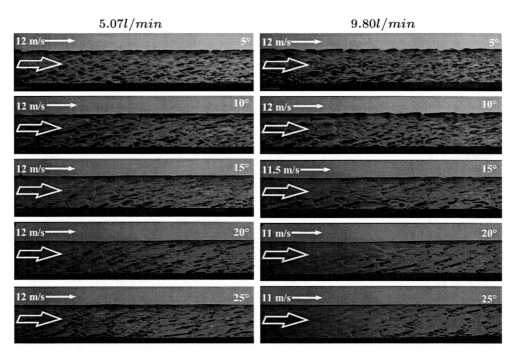

Fig. 4.14b: Influence of high viscosity on co-current waveforms under gravity

4.2.3.3   Combined Influence of Surface Tension and Viscosity

The left plot in Fig. 4.15 indicates that the liquid film with the superposition of low surface tension and high viscosity has no considerable difference from the reference case. The fact that the film loadings now incur similar pressure drops i.e. higher than the case with only low surface tension (see section 4.2.3.1) and smaller than the case with only high viscosity (see section 4.2.3.2) can easily be explained by looking at the distribution of the momentum transfer shown in Fig. 4.6 and the individual influence of the properties. The film is relatively thicker due to the action of increased viscosity but at the same time the film surface is relatively flat (less wavy, compare Fig. 4.16a with Fig. 4.14a) due to the decreased surface tension.

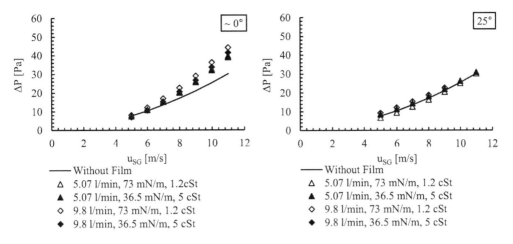

Fig. 4.15: Influence of high viscosity on pressure drops

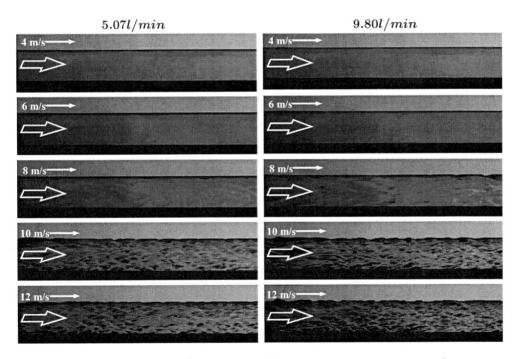

Fig. 4.16a: Influence of low surface tension and high viscosity on co-current waveforms

The supplementary influence of gravity on the quantitative (Fig. 4.15, right) and qualitative (see Fig. 4.16b) results is comparable to the reference case.

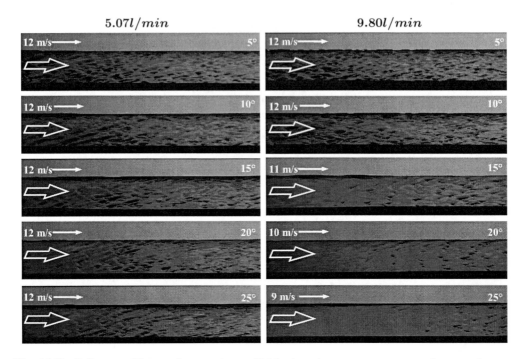

Fig. 4.16b: Influence of low surface tension and high viscosity on co-current waveforms under
gravity

4.2.4   Counter-Current Flow Regime

In this flow regime, the driving forces act in opposite directions (see Fig. 2.18). All counter-
current pressure drop measurements are performed using a ramp arrangement, $1mm$ thick,
$10mm$ high (from the channel floor) and extends $20mm$ into the test section, as shown by a
bold black line in Fig. 4.17. The sharp edge junction between the test section and the separator
tank, shown by a light grey filled circle in Fig. 4.17, leads to film roll ups and premature droplet
shedding. The ramp arrangement is used to assist a smooth gas liquid interaction in the test section.

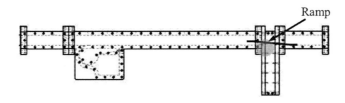

Fig. 4.17: Ramp arrangement for counter-current investigations

The shear stress acting on the film surface results in an almost immediate flow reversal, quickly followed by the supercritical film conditions (droplet shedding). This happens for an air velocity as low as $6m/s$, therefore, the influence of film loadings on pressure drops is studied with the help of the smallest possible inclination angle $(5°)$, see Fig. 4.18.

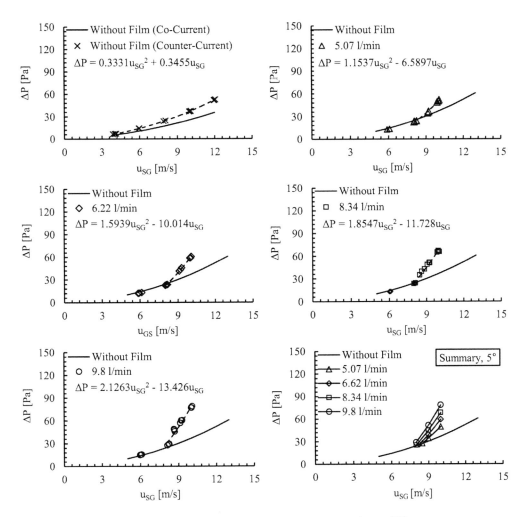

Fig. 4.18: Influence of film loading on pressure drops $(5°)$

Up to a certain air velocity, the film behaves in a similar manner as under pure gravity and almost no additional pressure drop can be measured. As mentioned in section 2.1.2.2, other researchers have also reported a lower limit on the critical gas velocity before the film may be disrupted and regarded this condition as a pre-flooding condition. Beyond this air velocity $(\sim 7 - 8m/s)$ a transition both in pressure drop and film behaviour takes place. Roy et al. (1989) and Vijayan et al. (2001) have also reported a similar trend in terms of pressure drop and film thickness just before the onset of flooding. The increase in film loading results in a shift of this transition point to a slightly lower gas velocity for all inclination angles (compare also film loading behaviour at $8m/s$ in Fig. 4.19). This is due to the fact that a relatively thicker film results in a higher local gas velocity, hence, a higher interfacial shear stress. This, as will be shown later, results in an early inception of surface instabilities and expected higher pressure drops.

Before going into the details of qualitative analysis, it should be noted that the counter-current film flow phenomena presented here are highly instantaneous and therefore, it is very difficult to deliver a universally valid description of the waveforms appearing on the film surface. This is the reason why no general description can be found in open literature as well. Therefore, the observations of other researchers can slightly vary from the description provided here. The description that follows is presented on the basis of careful observations of the flow phenomena in combination with the quantitative analysis and is supported by the corresponding pictures provided in Fig. 4.19.

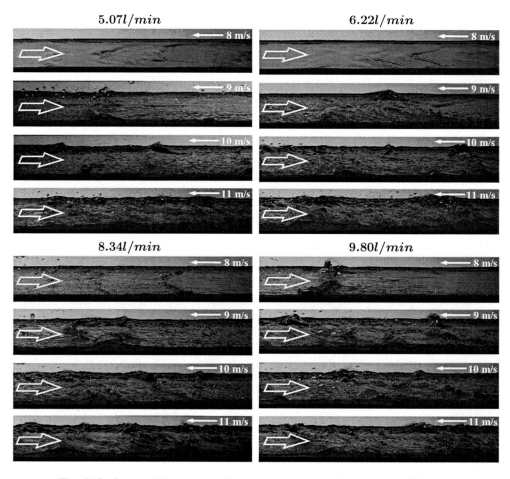

Fig. 4.19: Counter-Current waveforms at minimum inclination angle $(5°)$

At $8m/s$, a distorted lateral streak occupying the complete width of the channel appears. The streak is periodic and between the two streaks the surface is mostly smooth for the two smaller film loadings. For the two higher film loadings, one of the streak turns into a standing wave of small amplitude which is continuously being washed away by the incoming liquid and appears again periodically. The film surface between the streaks is covered with small ripples. At $9m/s$ periodic solitary waves, referred to as mini-slugs in several past studies e.g. Yamaguchi et al. (1982), Ghiaasiaan et al. (1997) etc...(see section 2.1.2.2 for details), arise from the streaks occupying the complete channel's width in a slightly distorted manner and display a to-and-fro motion (travelled couple of centimetres downstream then returned to the same position) on the film surface. This is due to the fact that the air flow is not strong enough to carry the thick wave along the channel length resulting in a recirculation zone on the film surface. The solitary waves are $0.1 - 0.15m$ apart and between the solitary waves the film surface is only occupied by small disturbances similar to the ripples at $8m/s$. Occasional droplet shedding from the tips of the solitary waves is also observed. At $10m/s$, the solitary waves, with a higher frequency than at $9m/s$, become non-uniformly distributed along the channel width. To-and-fro motion stops and the waves begin to travel down along the channel length. The film surface between the consecutive solitary waves is reduced to $0.05 - 0.1m$ and is occupied by variable amplitude 3D waves. Occasional but continuous droplet shedding is also observed in this category. At $11m/s$ and beyond, the non-uniformity among the solitary waves along the channel width keeps on increasing. The wave amplitude and frequency continuously increases with a continuous decrease in film surface between the consecutive solitary waves and a continuous but low rate droplet shedding is observed.

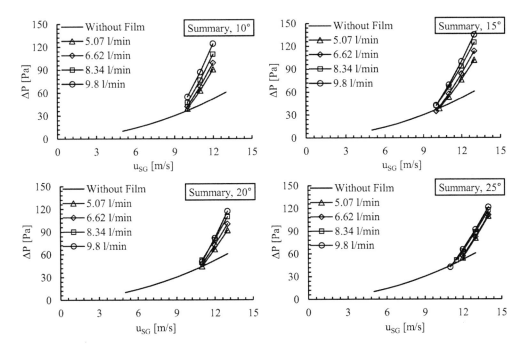

Fig. 4.20: Influence of film loading on pressure drops $(10 - 25°)$

From the preceding discussion it appears that the transition behaviour is dependent on the film loading, as is obvious for 5° i.e. a higher loading means an earlier transition. However, for large inclination angles this is not the case indicating that gravity plays a vital role in the determination of the transition point. Accordingly, it can be seen in Fig. 4.20 that from the critical angle of 15° (constant film dynamics hold), the transition points for different film loadings lie very close to each other. Although the curves exhibit different slopes due to different film momentums that has to be overcome by the counteracting air flow; yet, they also tend to overlap with an increasing influence of gravity (see Fig. 4.20 - 25°).

To further investigate the influence of gravity, the pressure drop of every investigated film loading is plotted against the inclination angles in Fig. 4.21. The transition point shifts towards higher air velocities for all investigated film loadings. Up to the critical inclination angle ($\sim$ 15°), this behaviour is quite plausible because the counter-current air flow has to overcome the gravity induced additional momentum. This behaviour, however, remains true even beyond the critical inclination angle as well. At first, this statement appears to be contradictory to the principle of constant film dynamics established earlier in this chapter. As it can be argued that since the counter-current air flow has to overcome always the same film momentum, the transition points should overlap after the critical inclination angle is reached. However, there is another point which requires consideration here. Although the film dynamics remain constant, every starting surface instability is subjected to higher gravitational force at higher inclination angles than at lower inclination angles.

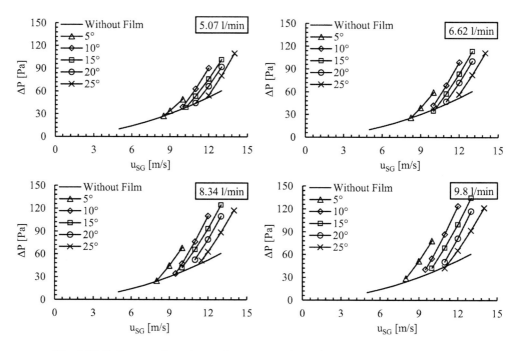

Fig. 4.21: Influence of gravity on the pressure drop of investigated film loadings

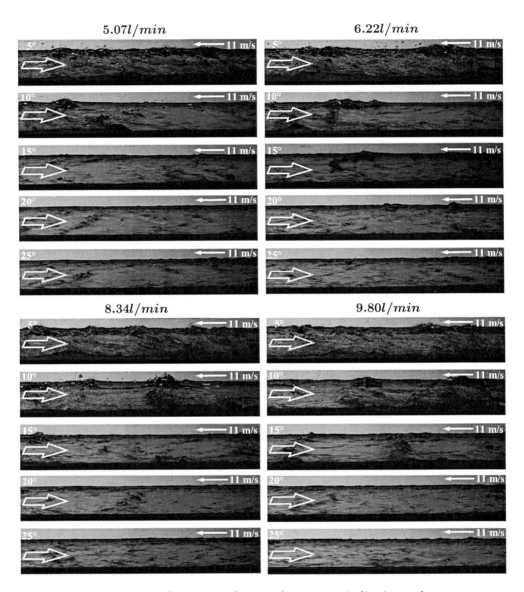

Fig. 4.22: Counter-Current waveforms with increasing inclination angles

In other words, the transition is delayed for higher inclination angles due to a higher gravitational influence on the first sustainable surface instability (wave). Once the instabilities are sustainable, the higher gravitational influence allows for only smaller wave amplitudes (as will be discussed later). Therefore, the film surface roughness reduces and results in a smaller pressure drop for the same conditions. This is exactly what is seen in Fig. 4.21 i.e. although the transition point shifts towards a higher air velocity for larger inclination angles, the pressure drop incurred at a lower angle for the same air velocity is much higher. As an example, compare the curves of 10° and

15° inclination angles in Fig. 4.21 at $12m/s$ air velocity. The high pressure drop incurred at 10° inclination as compared to 15° is attributed to the early transition (the first sustainable instability starts early) from flat to a wavy surface. The waves then grow with increasing air velocities, hence, higher pressure drops. The described behaviour of film dynamics with increasing inclination angles can easily be recognized with the help of the pictures provided in Fig. 4.22, where with every increase in the inclination angle the wave amplitude diminishes and this remains true for every investigated film loading.

4.2.4.1 Influence of Surface Tension

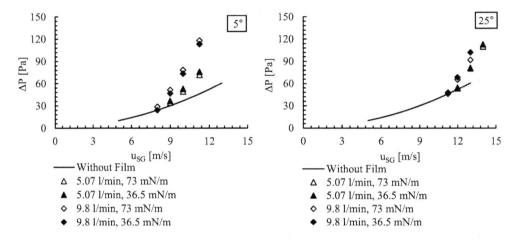

Fig. 4.23: Influence of low surface tension on pressure drops

The decrease in surface tension results in similar transition points and pressure drops for the investigated film loadings. From the visual observations presented in Fig. 4.24a, it can be seen that unlike the co-current case, in the counter-current regime the action of reduced surface tension cannot suppress the wave generation mechanism. Therefore, typical waveforms, as learned from the reference case, result here as well and correspondingly similar pressure drops are evident. The only noticeable difference to the reference case, which can also be observed by inspecting the pictures provided in Fig. 4.24a, is slightly more droplet shedding from the wave peaks. With increasing influence of gravity, again no difference can be observed in the pressure drop data for all the investigated film loadings (see Fig. 4.23, right plot). This can also be confirmed visually with the help of the film behaviour presented in Fig. 4.24b where for all inclination angles, the waveforms maintain similar amplitudes to those of the reference case (Fig. 4.22). Therefore, it can be concluded that in the counter-current regime, the reduced surface tension only plays a role in the early droplet shedding mechanism which can easily be explained by an increase $(2\times)$ in the Weber number. Ghiaasiaan et al. (1997) also investigated the effect of surface tension on the counter-current flows and did not find any significant influence on the behaviour of the gas-liquid interface.

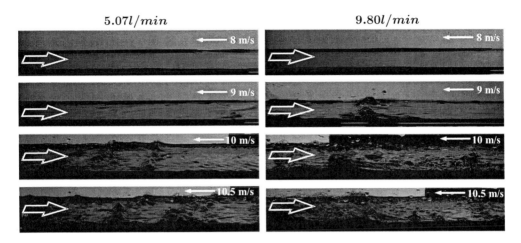

Fig. 4.24a: Influence of low surface tension on counter-current waveforms

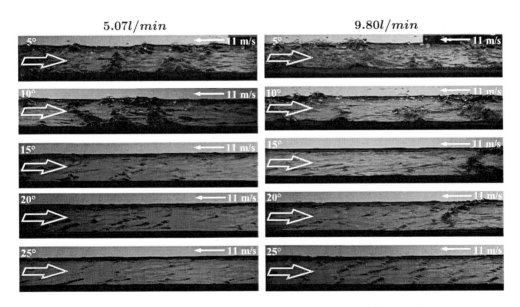

Fig. 4.24b: Influence of low surface tension on counter-current waveforms under gravity

4.2.4.2   Influence of Viscosity

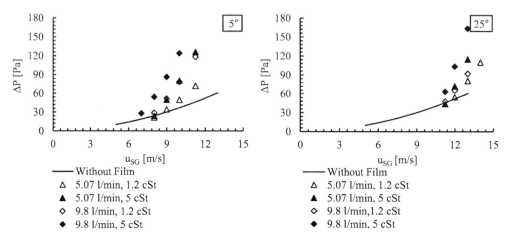

Fig. 4.25: Influence of high viscosity on pressure drops

With an increase in viscosity, the transition points shift towards lower gas velocities and the pressure drops increase for the investigated film loadings (see Fig. 4.25, left). As stated earlier, the film thickness increases with an increase in viscosity decreasing the air flow cross-sectional area; hence, higher shear on the film surface. This explains the early transition from flat to a wavy surface as compared to the reference case. These waves then grow with increasing air flow resulting in even higher amplitude waves, as shown in Fig. 4.26a, compared to the reference case (see Fig. 4.19). This is the reason for a relatively high pressure drop and a high rate droplet shedding as compared to the reference case.

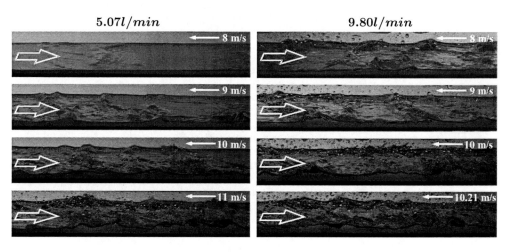

Fig. 4.26a: Influence of high viscosity on counter-current waveforms

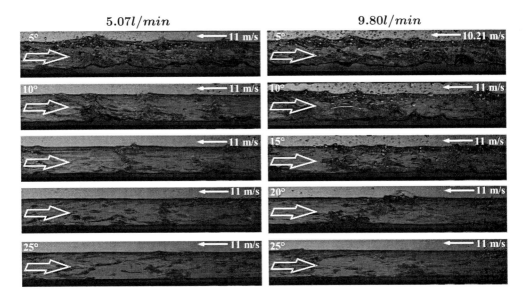

Fig. 4.26b: Influence of high viscosity on counter-current waveforms under gravity

Increasing the inclination angle, in contrary to the reference case, results in a similar pressure drop behaviour (see Fig. 4.25, right) to that at 5°. This can easily be explained based on the relatively high surface roughness due to the presence of interface waves, as show in Fig. 4.26b, which are absent in the reference case for the same conditions.

4.2.4.3 Combined Influence of Surface Tension and Viscosity

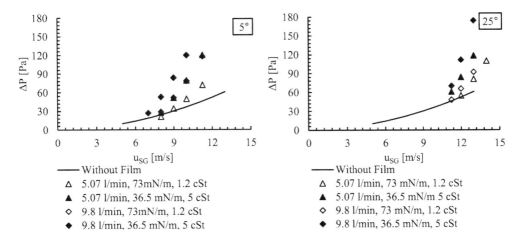

Fig. 4.27: Influence of low surface tension and high viscosity on pressure drops

The plots in Fig. 4.27 show a similar trend to the case with only high viscosity (section 4.2.4.2) indicating the dominance of viscosity on the pressure drops in the counter-current flow regime.

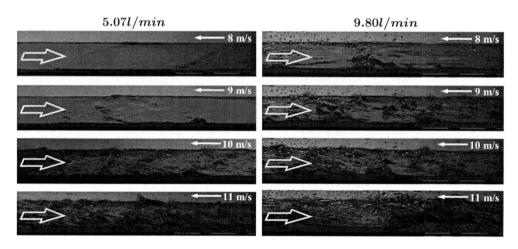

Fig. 4.28a: Influence of low surface tension and high viscosity on counter-current waveforms

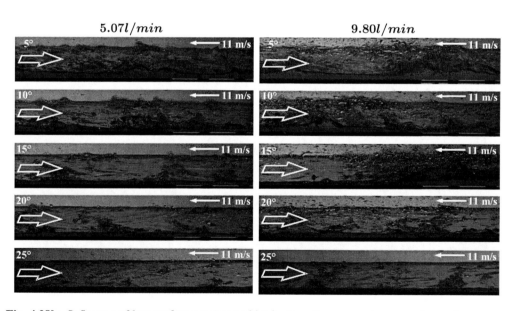

Fig. 4.28b: Influence of low surface tension and high viscosity on counter-current waveforms under gravity

The droplet generation in this case is, however, more pronounced than either in the reference case or the cases with individual change in properties. For the maximum investigated film loading, high rate droplet shedding is also evident even under the influence of gravity (see Fig. 4.28b). In sections 4.2.4.1 - 4.2.4.2, it was established that either lowering the surface tension or increasing the viscosity both results in a relatively high droplet shedding rate as compared to the reference case. Therefore, it can be said that the combined influence of reduced surface tension and high viscosity has enhanced the droplet generation phenomenon as evident here.

4.3 Significance of the Generic Results in a Bearing Chamber Environment

From the investigations carried out in this thesis, some simple rules for the momentum transfer to the film and transition from flat to a wavy film surface in the co- as well as the counter-current flow regime can be given. For thicker films, momentum transfer is higher and transition is earlier than for thinner films where comparatively lower momentum transfer and late transition is observed. Now extending the co-/counter-current findings to the bearing chamber flows, the investigations conducted in this thesis reveal that the bearing chamber air flow experiences almost no resistance on the co-current side. Furthermore, the oil film on the co-current side, as discussed earlier can be considered as laminar/nearly laminar and supercritical. The experiments conducted by Glahn et al. (1996) in the ITS model bearing chamber test rig also confirm these finding where none of the measured (film) velocity profiles could be found as fully turbulent. Moreover, increasing the shaft speed (increasing the interfacial shear) shows a clear tendency towards a laminar velocity profile. The Froude number calculated from the film velocity profile data reported by Glahn et al. (1996) (see Fig. 2.5) also shows that the film was supercritical for all boundary conditions. Another significance of the presented investigations is that the film dynamics in the horizontal case are not representative of the inclined cases. After a critical inclination angle, a transition from thick to thin $(viscous\ forces \approx gravitational\ forces)$ film can be theoretically proved. Accordingly, the film surface waves are largely damped in the inclined configuration. The roll waves are usually responsible for the droplet generation from thin films; however, they require much higher air velocities to appear on thin films than on thick (see e.g. Cohen and Hanratty (1961)). Therefore, it can be said that for the maximum possible mean core (air) velocity (see Gorse et al. (2003)) experienced by the film in the ITS HSBC test rig, droplet generation is very unlikely to appear on the co-current side.

On the other hand, the bearing chamber air flow experiences a major resistance in the counter-current half of bearing chambers. High enough air velocities will always result in solitary waves for all inclination angles. Considering the ITS HSBC test rig (see Fig. 2.1), where according to Gorse et al. (2003) a mean velocity of approx. 15 and $10m/s$ (while simulating high and intermediate pressure bearing chambers) can be calculated, a recirculating region with a high rate droplet shedding and only small surface instabilities with an occasional droplet shedding can be predicted for high and intermediate pressure bearing chambers (see Fig. 4.28b). Further increase in air velocity beyond the investigated range will soon result in a complete flow reversal accompanied with a high rate droplet shedding. After the transition point, the film flow can be considered as turbulent and sub-critical. The film is regarded as turbulent according to the general definition of turbulent flows i.e. transverse exchange of momentum; in other words, high mixing in the recirculation region formed by the solitary wave. From the experiments conducted using the ITS HSBC test rig e.g. Schmälzle (1997), Lipp (2003) etc...indeed an unstable recirculation region with droplet shedding and intense mixing was identified on the counter-current side. The sub-critical nature of the film can be defended on the basis of a continuous reduction in mean velocity and a continuous increase in thickness. This super- to sub-critical transition before a partial flow reversal (formation of a recirculation region) occurs on the film surface is explained by Gargallo et al. (2005) and Stäbler et al. (2006) in more detail.

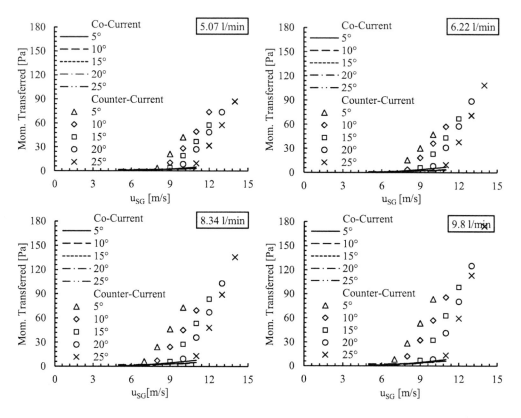

Fig. 4.29: Co-/Counter-Current momentum transfer under engine relevant conditions

Flouros (2005) used a realistic bearing chamber test rig and reported that increasing the oil flow rate resulted in a continuous increase in power consumption for all boundary conditions. This behaviour can be directly traced to the experimental investigations of this thesis where the momentum transferred to the film increases with an increase in film loading. Regarding the power consumption in bearing chambers, it can be said that the oil film incurs minimal losses on the co-current side as compared to the counter-current side (after the transition point). As can be seen from Fig. 4.29, the air loses multiple orders of magnitude more energy in the counter-current regime than in the co-current regime. This is due to the fact that in the counter-current regime air works against gravity whereas in the co-current regime gravity assists the shearing air flow.

From the discussion, it is also apparent that the churning losses in bearing chambers as a result of air/oil interactions must also be much larger on the counter-current side than on the co-current side.

4.4 Characterization of Bearing Chamber Typical Oil Film Flows

Based on the experimental investigations carried out in this thesis, some plausible characteristics of the oil film dynamics near scavenge off-takes can now be identified. Table 4.5 summarizes the characteristics of a film flow that may result in a bearing chamber in the absence of coexisting flow phenomena (e.g. droplet interaction, off-take disturbances etc...).

Co-Current	*Counter-Current*
• Flat surface or shallow waves	• Big solitary waves
•Very low probability of droplet shedding	• Droplet shedding
• Laminar/nearly laminar	• Turbulent
• Supercritical	• Sub-critical
• Can be considered as fully developed	• Definitely undeveloped
• Almost no churning loss	• High churning losses

Table 4.5: Film characterization in a bearing chamber environment

5 Modelling of Shear Driven Thick Film Flows

As mentioned earlier in chapters two and three, the experimental investigations in model bearing chamber test rigs are limited by a number of factors and the knowledge gained cannot be concluded in generalized design rules. This fact points towards the necessity of alternate design tools which can be employed to study the bearing chamber flow phenomena more efficiently and comprehensively. This chapter therefore, investigates the potential offered by the state of the art numerical techniques for simulating the oil film flow in the vicinity of the scavenge off-takes. Based on the important physical features of the oil film, multiphase models are investigated for their applicability. The choice of the most suitable model is made on the basis of accuracy and efficiency (time & cost). The model is then theoretically and practically investigated and a serious shortcoming is ascertained. Two methods consisting of a simplified and a more sophisticated approach are introduced to improve the model performance significantly.

5.1 Multiphase Modelling Techniques

In the last chapters, it is learned that the major features of the bearing chamber typical oil film flow are governed by the interaction between the oil film and the driving forces. It is important to know the physical features to be modelled because all multiphase flow models are flow regime dependent. A flow regime can be defined as a characteristic phase distribution, recognized by a certain geometric arrangement of phases and is clearly observable on the axial and lateral cross-sections of the transporting conduits e.g. pipe or channel. Based on the experimental investigations conducted in the last chapter, a wavy stratified flow with a well-defined but highly dynamic interface separating the liquid from the gas phase is observed for all relevant conditions. The highly dynamic nature of the interface poses a major difficulty towards a mathematical formulation describing the behaviour of a wavy stratified flow. Apart from this, the physical properties of the flow field vary considerably across the interface and must be considered as discontinuities; otherwise, the flow features cannot be resolved accurately. Despite the fact that surface waves appear to follow a certain pattern, quantification in time and space is very difficult. As mentioned above, due to a complex and highly dynamic structure of the multiphase flows, it is very difficult to find a general mathematical formulation applicable to all multiphase flow regimes. A general mathematical formulation is also not feasible because the modelling of additional terms arising from the interphasic interactions, due to a lack of thorough physical understanding, offers a very little or no improvement on the expense of an exponential increase in the computational resources. However, due to their engineering and scientific importance, a number of mathematical formulations dealing with individual flow regimes have been proposed in literature and a continuous development in this regard have led to fairly accurate (case dependent) solutions.

The basic idea behind all mathematical formulations is derived from the concept of a Two-Fluid Model which treats the different phases as interpenetrating continua. At every location in a flow domain individual conservation equations exist for every phase. The phases interact with each other and the local flow field conditions result from these interphasic interactions. The interactions are realized by means of introducing interface exchange terms in the conservation equations of each phase, hence, establishing a link between the individual conservation equations. In general, there are two approaches to model all multiphase flows, the all continuum and the discrete phase approach where all secondary phases (for low volume fractions $< 10-12\%$) are arbitrarily dispersed in the continuous primary phase (FLUENT 6.3 User's Guide). In the case of a bearing chamber film flow i.e. a wavy stratified flow, only the all continuum approach is appropriate.

5.1.1 Two-Fluid Model

In literature especially the commercial CFD codes refer to the Two-Fluid Model as the full Eulerian, Euler-Euler or the inhomogeneous modelling approach. The general conservation equations can be derived by introducing the concept of a phasic volume fraction 'α'. This concept suggests that the volume of a phase cannot be occupied by the other phases. From this consideration the multiphase Navier-Stokes equations can be derived directly from the single phase Navier-Stokes equations by defining a phase indicator function 'γ_q' of a phase 'q' which is assumed to be a continuous function in space and time. The indicator function takes the value of 1 if the space under consideration is occupied by the phase 'q' and the value 0 in the absence of phase 'q'. Performing a time-average (Reynolds-Averaging) of the phase indicator function over the time interval 'Δt', the volume fraction 'α_q' of the phase 'q' at a point in space according to Drew and Passman (1999) can be written as follows

$$\alpha_q = \overline{\gamma}_q = \frac{1}{\Delta t} \int_t^{t+\Delta t} \gamma_q(t) \mathrm{d}t \tag{5.1}$$

Using this definition of the phasic volume fraction, the volume 'V_q' occupied by the phase 'q' is given by

$$V_q = \int \alpha_q \mathrm{d}V \; where \; \sum_{q=1}^{n} \alpha_q = 1 \tag{5.2}$$

The effective density '$\hat{\rho}_q$' of phase 'q' can now be written as

$$\hat{\rho}_q = \alpha_q \rho_q \tag{5.3}$$

For a two-phase system consisting of a gas 'G' and a liquid 'L' (*i.e.* $q = G,L$), the volume fractions of the phases are coupled according to the relationship

$$\alpha_G + \alpha_L = 1 \tag{5.4}$$

To describe every phase of a multiphase flow as continuum, the flow field variables '$\overline{\phi}_q$' of every phase, where '$\overline{\phi}_q$' can be a scalar or vector variable, must be time-averaged in a similar manner as done for the phase indicator function. The phase-averaged flow variables at a point in a flow domain can now be written as

$$\overline{\phi}_q = \frac{1}{\alpha_q} \int_t^{t+\Delta t} \gamma_q(t) \phi_q \mathrm{d}t \tag{5.5}$$

By applying the concept of phase-average on the single phase mass conservation equation, the Two-Fluid mass conservation equation for a multiphase system can be written as

$$\frac{\partial \hat{\rho}_q}{\partial t} + \frac{\partial (\hat{\rho}_q \overline{u}_i^q)}{\partial x_i} = \Gamma_q \qquad (5.6)$$

With the help of 'Γ_q' term in the continuity equation, the phase change phenomena like evaporation/condensation or an explicit mass source/sink can be modelled. The general implementation of 'Γ_q' in commercial CFD codes is given in equation 5.7.

$$\Gamma_q = (\dot{m}_{LG} - \dot{m}_{GL}) + S_q \qquad (5.7)$$

In Eq. 5.7, $\dot{m}_{LG}$ characterizes the mass transfer from the liquid phase 'L' to the gas phase 'G' (e.g. evaporation, boiling etc.) and $\dot{m}_{GL}$ characterizes the mass transfer from the gas phase 'G' to the liquid phase 'L' (e.g. condensation, liquefaction etc). The mechanisms of mass transfer in all commercial codes can be specified separately. The term 'S_q' offers the possibility of specifying an additional, independent mass source(s) in the domain.

Similar to the continuity equation, applying the phase-average on the single phase Navier-Stokes equation, the Two-Fluid turbulent momentum conservation equation results which can be written as follows

$$\frac{\partial (\hat{\rho}_q \overline{u}_i^q)}{\partial t} + \frac{\partial (\hat{\rho}_q \overline{u}_i^q \overline{u}_j^q)}{\partial x_j} = -\alpha_q \frac{\partial P}{\partial x_i} + \frac{\partial}{\partial x_j} \left[\alpha_q \left(\overline{\tau}_{ij}^q + \overline{\tau}_{ij}^{Re,q} \right) \right] + \hat{\rho}_q g_i + F_{q,i} + M_q^{(\Gamma_q)} \qquad (5.8)$$

In Eq. 5.8, 'P' is the pressure and is shared by all phases. The modelling of pressure nonequilibrium is much more complex (Ishii (1975)). Omgba-Essama (2004) reviewed in his work that the pressure difference between two phases is caused by three main effects: surface energy of a curved interface, mass transfer and dynamic effects. In the first case the simple existence of an interface (probably curved) requires that some pressure difference exists between the phases. This pressure difference is proportional to the interfacial surface tension and inversely proportional to the radius of curvature and is usually neglected in most applications. The second effect is noticeable when the mass flux due to phase change is large at the interface between the phases; e.g., large evaporation or condensation rates. The final effect occurs when one phase has a larger pressure relative to the other phase due to very rapid energy deposition or pressurization effects. For most engineering applications including wavy stratified film flows, the pressure difference between the phases can be neglected and an equal pressure assumption gives reasonably accurate results. However, if the pressure difference cannot be neglected a local constitutive relation must be introduced accounting for the pressure difference between the phases.

The effect of turbulence on the flow field is modelled with the help of Reynolds stresses, given by the term '$\overline{\tau}_{ij}^{Re,q}$' in Eq. 5.8, of the individual phases. The Reynolds stresses result from fluctuating velocity fields, causing an intense mixing of transported quantities like momentum, energy and species concentration and thereby also fluctuations in the transported quantities. A direct simulation of the small scales and high frequency fluctuations are computationally too expensive to be feasible for a practical engineering purpose. For this reason, a set of less expensive governing equations are obtained by using a time-averaged (Reynolds averaging) approximation of the instantaneous governing equations which removes the small scales. In the new set of equations additional terms arise which are modelled using a turbulence model. Of the classical models, the eddy viscosity turbulence models are presently by far the most widely used and validated. They are based on the presumption that there exists an analogy between the action of viscous stresses and Reynolds stresses on the mean flow (the Boussinesq Assumption). Both stresses appear on the right hand side of the momentum equation and in Newton's law of viscosity the viscous stresses are taken to be proportional to the rate of deformation of fluid elements (see Versteeg and Malalasekera (1995) for details). Therefore, the laminar and turbulent stresses in the momentum balance equation (Eq. 5.8) for the phase 'q' are modelled as

$$\frac{\partial}{\partial x_j}\left[\alpha_q\left(\overline{\tau}_{ij}^q + \overline{\tau}_{ij}^{Re,q}\right)\right] = \frac{\partial}{\partial x_j}\left[\alpha_q\mu_{q,eff}\left(\frac{\partial \overline{u}_i^q}{\partial x_j}\right)\right] \; with \; \mu_{q,eff} = \mu_q + \mu_t \qquad (5.9)$$

'μ_t' is called the 'turbulent viscosity'. It is a virtual parameter and is used to model the effect of turbulence on the flow field. A detailed explanation follows in the later section dealing with turbulence modelling.

'g_i' is the component of the gravitational vector and '$F_{q,i}$' are the forces defining interphasic interactions between the phases e.g. interfacial shear stress. These forces depend on the nature of a multiphase flow field i.e. flow regime dependent. The most general form of an interaction term results from the velocity difference between the phases and is also used in commercial codes (see FLUENT 6.3 User's Guide for details).

$$F_{q,i} = K_{LG,i}(\vec{u}_G - \vec{u}_L) \; where \; K_{LG,i} = f(\alpha_L, \alpha_G, f_i, \dots) \; \& \; f_i = f(C_D, Re) \qquad (5.10)$$

Nearly all definitions of '$K_{LG,i}$' include a drag fraction 'f_i' that is based on the drag coefficient (C_D) and the relative Reynolds number (Re). It is this drag function that differs among the different exchange-coefficient models. For all these situations,'$K_{LG,i}$' should tend to zero whenever one of the phases is not present within the domain. To enforce this, the drag function 'f_i' is always multiplied by the volume fraction of the phases.

'$M_q^{(\Gamma_q)}$' is the momentum source term due to mass transfer across the interface and is given by

$$M_q^{(\Gamma_q)} = (\dot{m}_{LG}\vec{u}_{LG} - \dot{m}_{GL}\vec{u}_{GL}) \qquad (5.11)$$

In Eq. 5.11, '$\vec{u}_{LG}$'is the interphase velocity and is equal to the liquid velocity '$\vec{u}_L$' if the mass of liquid phase is being transferred to the gas phase and vice versa.

The commercial codes also offer the possibility of modelling other forces like external body forces (e.g. due to a magnetic field), lift forces (e.g. on particles in continuous media due to velocity gradients or density differences) and virtual mass forces that occur when a disperse phase accelerates relative to the continuous phase. The inertia of the continuous phase mass encountered by the accelerating particles (e.g. droplets or bubbles, provided the slip is large enough) exerts a "virtual mass force" on the particles. In addition to that an infinite number of other momentum sources can be introduced in the momentum balance equation.

Omgba-Essama (2004) reviewed in his work that the temperature difference between the phases is fundamentally induced by the time lag of energy transfer between the phases at the interface as thermal equilibrium is reached. If a multiphase system involves rapidly changing flow conditions due to area changes in steady flow or transient conditions then the time lag for reaching thermal equilibrium between the phases may become significant in comparison to the characteristic time it takes for flow conditions to change. According to Corradini (1997), this condition may be estimated by computing a characteristic Fourier number, given by the ratio of the current time to the steady state time, for the system under the expected flow conditions. Hence, when thermal non-equilibrium becomes important, the possibility of a temperature difference must be included by separate energy balances in a multiphase model. Since the research carried out in this report deals with isothermal flow conditions, the energy equation of the Two-Fluid model is not presented here. The reader can consult Corradini (1997) or Drew and Passman (1999) for details. It should be noted that when a Two-Fluid (full Eulerian) model is used, a number of interfacial transport coefficients e.g. '$K_{LG,i}$' are defined and require constitutive relations to close the overall model (Corradini (1997)). This approach has an advantage in that the actual transport processes can be rigorously defined; however, the disadvantage is that it is required to model these kinetic processes in detail, which implies a much greater depth of experimental data and insight (Omgba-Essama (2004)). As mentioned in Corradini (1997), the constitutive relations for interfacial transfer are complicated functions of the fluid velocities and their local properties. These kinetic models are also a strong function of the multiphase flow pattern. For example, the models developed for the interfacial shear stress or heat flux will be significantly different for a dispersed flow pattern in contrast to a wavy stratified flow pattern. In fact, the interfacial models would be different if there were gas bubbles in a liquid versus liquid droplets in a gas.

5.1.2   Homogenous Model

The (wavy) stratified flow of two immiscible fluids where interface (free surface) tracking is of utmost importance can be modelled using a single phase or homogenous approximation, commonly known as the Free Surface/VOF (Volume-Of-Fluid) Model, of the Two-Fluid Model. In this model, it is assumed that the velocity, temperature and pressure of the phases are equal at the interface. This assumption is based on the condition that the differences in these three variables (and chemical potential if chemical reactions are considered) will promote momentum, energy, and mass transfer between the phases rapidly enough so that thermodynamic equilibrium is reached. This will be the case for drag-dominated flows where the two phases are strongly coupled and their relative velocities equalise over short spatial length scales. This is also valid when one phase is finely dispersed in another phase, a large interfacial area is generated and, under certain circumstances, this assumption can be made; e.g., bubbly flow of air in water or steam in water at high pressures (Omgba-Essama (2004)).

The governing equations of the VOF model, similar to the Two-Fluid Model, are derived by introducing the concept of phase indicator function (see Eq. 5.1) in the single phase Euler equations (see Hirt and Nichols (1981) for details). The interface between the phases is not sharp but smeared over several cells and the phase indicator function takes the value between zero and one as shown in Fig. 5.1. The tracking of interface(s) between the phases is accomplished by solving an additional advective transport equation for the volume fraction of one (or more) of the

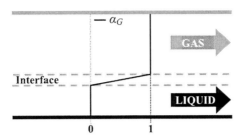

Fig. 5.1: Concept of a smeared interface

phases. For the 'q^{th}' phase, this equation has the following form

$$\frac{\partial \alpha_q}{\partial t} + \frac{\partial \left(\alpha_q \overline{u}_i^q\right)}{\partial x_j} = 0 \qquad (5.12)$$

The volume fraction of the primary phase will be computed based on the following constraint

$$\sum \alpha_q = 1 \qquad (5.13)$$

The discontinuous physical properties (density and viscosity) across the interface, analogous to the phase indicator function, are continuous functions in time and space in the governing equations. This continuity in physical properties across the interface is achieved by the volume fraction based averaging in the cells occupied by the interface which is as follows

$$\rho = \alpha_G \rho_G + \alpha_L \rho_L \;\rightarrow\; \rho = \sum \alpha_q \rho_q \qquad (5.14)$$

$$\mu = \alpha_G \mu_G + \alpha_L \mu_L \;\rightarrow\; \mu = \sum \alpha_q \mu_q \qquad (5.15)$$

Using the concept of phase indicator function, the single phase approximation of the Two-Fluid governing equations can now be written as

$$\frac{\partial \rho}{\partial t} + \frac{\partial \left(\rho \overline{u}_i\right)}{\partial x_i} = \Gamma_q \qquad (5.16)$$

$$\frac{\partial \left(\rho \overline{u}_i\right)}{\partial t} + \frac{\partial \left(\rho \overline{u}_i \overline{u}_j\right)}{\partial x_j} = -\frac{\partial P}{\partial x_i} + \frac{\partial}{\partial x_j}\left(\overline{\tau}_{ij} + \overline{\tau}_{ij}^{Re}\right) + \rho g_i + F_i \qquad (5.17)$$

'Γ_q' in the continuity equation (Eq. 5.16) has the usual meaning as given by Eq. 5.7. All terms in the momentum equation (Eq. 5.17) have also the usual meanings as already discussed in section 5.1.1. With the help of 'F_i', external body forces can be modelled. The dependence on the volume fractions of the phases is realized through the physical properties 'ρ' and 'μ' of the fluids. Due to the inherent nature of the homogenous modelling technique, interphasic forces e.g. shear stress do not need to be modelled explicitly as in the Two-Fluid momentum equation but rather result from the velocity gradient at the interface. However, one limitation of the shared field approximation is that in cases where a large velocity difference exists between the phases, the accuracy of the velocities computed near the interface can be adversely affected (FLUENT 6.3 User's Guide). This point will be further discussed later in sections 5.2 - 5.3 where the model feasibility analysis is dealt.

5.1.2.1 Surface Tension and Wall Adhesion Modelling

For wavy stratified flows, the surface tension effects need to be included in the model. Nearly all free surface model implementations in commercial codes make use of the continuum surface force (CSF) concept proposed by Brackbill et al. (1992) in the early 90's. This concept models the surface tension force as a volume force concentrated at the interface, rather than a surface force; hence, can be added in the momentum equation just like any ordinary source term. For a liquid and gas system as is investigated in this thesis, the surface tension force given by the continuum surface force model is modelled as

$$\vec{F}_{LG} = \vec{f}_{LG} \delta_{LG} \tag{5.18}$$

$$where \ \vec{f}_{LG} = -\sigma_{LG} \kappa_{LG} \vec{n}_{LG} + \vec{\nabla}_s \sigma \tag{5.19}$$

$$\delta_{LG} = |\vec{\nabla} \alpha_{LG}| \tag{5.20}$$

'σ' is the surface tension, '$\vec{n}_{LG}$' is the interface normal vector pointing from the heavy fluid (Liquid) to the lighter fluid (Gas) calculated from the gradient of a smoothed volume fraction, '∇_s' is the gradient operator on the interface and 'κ_{LG}' is the surface curvature defined by

$$\kappa_{LG} = \vec{\nabla} \cdot \vec{n}_{LG} \tag{5.21}$$

The two terms summed on the right hand side of Eq. 5.19 reflect the normal and tangential-components of the surface tension force, respectively. The normal component arises from the interface curvature and the tangential component from variations in the surface tension due to the Marangoni effect. The 'δ_{LG}' term is often called the interface delta function; it is zero away from the interface, thereby, ensuring that the surface tension force is active only in the cells near the interface.

The VOF model also offers the possibility to account for the wall adhesion by specifying the contact angle 'θ', where 'θ' is the angle that the interface makes with a wall through the heavy fluid. This may be necessary where the climbing of a fluid on a wall is important, in other words, the wall affects the film flow and needs to be simulated. If wall adhesion is simulated, consistency is ensured by the constraint that the interface normal vector used for the calculations of both curvature and the surface tension force must satisfy the wall contact angle.

5.1.3 Model Selection

At present, the inhomogeneous modelling (full Eulerian and its derivative the Mixture Model) approach is limited to the disperse flow regimes where each secondary phase is assumed to constitute of spherical droplets, bubbles or particles. Although some commercial codes do support the interface tracking algorithm in combination with the Eulerian model e.g. ANSYS CFX; however, the interphasic interactions have to be defined by the user. In the case of shear driven wall films, this means that the interface shear stress needs to be modelled by the user. Unlike a disperse phase where a fixed geometry can be assumed for the droplets, bubbles or particles and therefore, a drag law can be derived; defining such a model for wavy stratified flows is in no way trivial. This is attributed to the highly dynamic interface between the phases with theoretically little prior knowledge as to what kind of interface geometry will evolve in time and space under the provided boundary conditions. It is this variable and uncertain interface geometry which is responsible for the lack of a successful empirical/analytical correlation for the drag law. As mentioned in detail in section 2.2.2, there is no universal drag law available which describes the momentum transfer distribution in time and space between the continuous phases especially for undeveloped flows. Former studies have mostly attempted to find a global description of the interactions at the interface (e.g. interface shear stress) either by modifying correlations used for single phase flows or by more detailed empirical correlations where two phase flow effects have been modelled using experimental data or rough wall analogies. However, all these approaches are limited to the fully developed flows. In engineering practice, the primary interest in simulating the wavy stratified flows is to gain insight on the flow pressure drop which is a function of the interfacial shear stress distribution. Therefore, the models which require a prior information on the shear stress distribution at the gas-liquid interface cannot fulfil the primary motivation to simulate the wavy stratified flows. On the other hand, the single phase approximation of the full Eulerian method, the Free Surface or the VOF model, does not require a prior knowledge of the shear stress distribution on the gas-liquid interface. Due to a single set of momentum equations, a single velocity field is shared by the phases; hence, the interfacial shear stress does not need to be explicitly specified instead it is embedded in the velocity gradient between the phases at the gas-liquid interface. This feature of the VOF Model together with half the number of equations to be solved makes it a potential tool for the simulation of bearing chamber typical film flows (wavy stratified flows) dealt in this work. However, as mentioned in section 5.1.2, one inherent characteristic of the shared-field approximation is that in cases where large velocity and physical properties gradients exist between the phases, the accuracy of the velocities computed near the interface can be adversely affected. A large velocity gradient across the interface is the main feature of shear driven liquid wall films. Therefore, it is very clear from the fundamentals of this model that this model was not developed for the shear driven wall films, i.e. where the entire driving energy of one phase is coming from another phase, rather than for pure interface tracking between the phases. The primary phase, usually the heavier of the two, is traced whereas the secondary phase is merely there to keep the VOF basic assumption, hence, solver mathematics true i.e. voids are not allowed. Nevertheless, since VOF is the only suitable choice for the shear driven wall films (see also section 2.2.2), in the following, it is thoroughly investigated for its capabilities, shortcomings and possible improvements.

5.2 Multiphase Turbulence Modelling

The main characteristics of turbulent multiphase flows, similar to the single phase, are fluctuating velocity fields. The Reynolds stresses of the individual phases '$\overline{\tau}_{ij}^{Re,q}$' are given by decomposing the velocity into its average and instantaneous value and then carrying out the similar phase averaging procedure on the convective terms as described in section 5.1.1. Therefore, the basic idea behind the multiphase turbulence modelling is the same as for the single phase with the difference that multiphase turbulence models are complemented with flow regime specific additional terms which can be interpreted as additional sources of turbulence production and its dissipation (Wintterle (2008)). Accordingly, in practice, the turbulence models used to model single phase flows are adapted to solve the multiphase problems as well. The most commonly used turbulence models, as mentioned earlier, are based on the experimental observation that turbulence decays unless there is shear in isothermal incompressible flows. Furthermore, turbulent stresses are found to increase as the mean rate of deformation increases. It was proposed by Boussinesq in 1877 that the Reynolds stresses could be linked to the mean rates of deformation as follows

$$\bar{\tau}_{ij}^{Re} = -\rho \overline{u_i' u_j'} = \mu_t \left(\frac{\partial U_i}{\partial x_j} + \frac{\partial U_j}{\partial x_i} \right) \tag{5.22}$$

Before going into the details of turbulence modelling, the different levels of approximations involved in turbulence modelling to close the momentum equations are briefly discussed in the following (for details see Versteeg and Malalasekera (1995)):

Algebraic models: An algebraic equation is used to compute the turbulent/eddy viscosity. The Reynolds stress tensor is then computed using the 'Boussinesq assumption' which relates the Reynolds stress tensor to the mean velocity gradients and the eddy viscosity (Eq. 5.22). Algebraic models are also called the mixing length or zero-equation turbulence models. In two-dimensional thin shear layers the changes in the flow direction are always so slow that the turbulence can adjust itself to local conditions. If the convection and diffusion of turbulence properties can be neglected it is possible to express the influence of turbulence on the mean flow in terms of a mixing length. If convection and diffusion are not negligible - as is the case for example in recirculating flows - a compact algebraic description for the mixing length is no longer feasible. The mixing length model lacks this kind of generality. The way forward is to consider statements regarding the dynamics of turbulence.

One-equation models: In one-equation models, a transport equation is solved for a turbulent quantity (usually the turbulent kinetic energy) and a second turbulent quantity (usually a turbulent length scale) is obtained from an algebraic expression. The turbulent viscosity is calculated from the 'Boussinesq assumption'.

Two-equation models: The two-equation models, e.g. '$k-\varepsilon$' and '$k-\omega$' models, focus on the mechanisms that affect the turbulent kinetic energy and are more general than either the algebraic or the one-equation turbulence models but are also more costly. The description of turbulence by two-equation models allows for the effects of transport of turbulence properties by the mean flow, diffusion and for the production and destruction of turbulence. Two transport equations, one for the turbulent kinetic energy 'k' and a further one for the rate of dissipation of turbulent kinetic energy 'ε' (or specific dissipation rate 'ω'), are solved. The eddy viscosity is computed from the two transported scalars and the Reynolds stress tensor is computed using the 'Boussinesq assumption'. The underlying assumption of both these models is that the turbulent viscosity 'μ_t' is isotropic. In other words, the ratio between Reynolds stress and mean rate of deformation is the same in all directions. This assumption fails in many categories of flows where it leads to inaccurate flow predictions. Here it is necessary to derive and solve transport equations for the Reynolds stresses themselves.

Reynolds stress models: An individual transport equation is derived for all Reynolds stresses. It may at first seem strange to think that a stress can be subjected to transport. However, it is only necessary to remember that the Reynolds stresses initially appeared on the left hand side of the momentum equations and are physically due to convective momentum exchanges as a consequence of turbulent velocity fluctuations. Fluid momentum - mean momentum as well as fluctuating momentum - can be transported by fluid particles and therefore the Reynolds stresses can also be transported. The six transport equations, one for each Reynolds stress, contain diffusion, pressure-strain and dissipation terms whose individual effects are unknown and cannot be measured. In Reynolds stress equation models (also known in the literature as second-order or second-moment closure models) assumptions are made about these unknown terms and the resulting PDE's are solved in conjunction with the transport equation for the rate of dissipation of turbulent kinetic energy 'ε'.

5.2.1   Selection of a Turbulence Model

Due to the reasons stated above, neither the zero-equation nor the one-equation model is sufficient to capture the turbulence dynamics of shear driven stratified flows. According to Versteeg and Malalasekera (1995), the design of Reynolds stress models is an area of vigorous research and the models have not been validated as widely as the two-equation models. A much more far-reaching set of modelling assumptions reduces the PDE's describing the Reynolds stress transport to algebraic equations to be solved alongside the 'k' and 'ε' equations of the '$k-\varepsilon$' model. This approach leads to the Algebraic stress models that are the most economical form of Reynolds stress models able to introduce anisotropic turbulence effects into the CFD simulations. However, solving for the seven extra PDE's gives rise to a substantial increase in the cost of CFD simulations when compared to the two-equation models. Besides, the lack of knowledge about the behaviour of turbulent quantities near the gas-liquid interface does not allow the application of either the Reynolds stress or the Algebraic stress models at this point. Therefore, the work conducted in this thesis employs two-equation turbulence models.

5.2.2   Two-Equation Turbulence Models

At first, the turbulence modelling concepts of a single phase are derived which are later extended to multiphase flows. This step is also necessary because the homogenous multiphase model (VOF), which is the preferred model in this research work, can only accept a homogenous turbulence model which is based on single phase turbulence modelling.
The transport equation for the turbulent kinetic energy can be derived by the multiplication of each of the instantaneous Navier-Stokes equations with the appropriate fluctuating velocity components. Addition of all the results and subtracting it from the equation of mean kinetic energy yields the equation for the turbulent kinetic energy 'k' (Tennekes and Lumley (1972)):

$$
\underbrace{\frac{\partial(\rho k)}{\partial t}}_{I} + \underbrace{\frac{\partial(\rho k U_i)}{\partial x_i}}_{II} = \frac{\partial}{\partial x_i}\left(\underbrace{-\overline{P'u_i'}}_{III} + \underbrace{2\mu\overline{u_i'e_{ij}'}}_{IV} - \underbrace{\rho\frac{1}{2}\overline{u_i'\cdot u_i'u_j'}}_{V} \right) - \underbrace{2\mu\overline{e_{ij}'\cdot e_{ij}'}}_{VI} - \underbrace{\rho\overline{u_i'u_j'}\cdot E_{ij}}_{VII}
$$

$$(5.23)$$

The term VII gives a positive contribution to Eq. 5.23 and represents a production term. This expresses mathematically the conversion of mean kinetic energy of the flow into turbulent kinetic energy. The viscous dissipation term VI gives a negative contribution to Eq. 5.23 and it is always the main destruction term in the turbulent kinetic energy equation, of a similar order of magnitude to the production term and never negligible.

In the standard '$k-\varepsilon$' model (Launder and Spalding (1974)), the production term VII is derived analogous to the Boussinesq assumption i.e. proportional to the mean value of the transported quantity called the gradient diffusion method. Prandtl number 'σ_k' connects the diffusivity 'Γ_k' of 'k' to the eddy viscosity '$\sigma_k = \frac{\mu_t}{\Gamma_t}$'. Since the turbulent transport of momentum and heat or mass is due to the same mechanism - eddy mixing - the value of the turbulent diffusivity can be assumed to be close to that of the turbulent viscosity. With these definitions, the transport equation for turbulent kinetic energy 'k' becomes

$$\frac{\partial(\rho k)}{\partial t} + \frac{\partial(\rho k U_i)}{\partial x_i} = \frac{\partial}{\partial x_i}\left[\left(\mu + \frac{\mu_t}{\sigma_k}\right)\frac{\partial k}{\partial x_i}\right] + 2\mu_t E_{ij} \cdot E_{ij} - \rho\varepsilon \qquad (5.24)$$

The pressure term III in Eq. 5.23 cannot be measured directly. Therefore, its effect is accounted for in Eq. 5.24 within the gradient diffusion term. *(Note: For detailed explanation of every term in Eq. 5.23 see Versteeg and Malalasekera (1995))*

It is possible to develop similar transport equations for all other turbulence quantities including the rate of viscous dissipation 'ε' (see Bradshaw et al. (1981)). The exact 'ε' equation, however, contains many unknown and immeasurable terms. Therefore, the standard '$k-\varepsilon$' model (Launder and Spalding (1974)) dissipation equation is based on the best understanding of the relevant processes causing changes to this variable. Accordingly, the transport equation for 'ε' assumes that its production and destruction terms are proportional to the production and destruction terms of 'k' (Eq. 5.24). Adoption of such forms ensures that 'ε' increases rapidly if 'k' increases rapidly and that it decreases sufficiently fast to avoid (non-physical) negative values of turbulent kinetic energy if 'k' decreases. However, an additional multiplier '$\frac{\varepsilon}{k}$' is required in the production and destruction terms to make these terms dimensionally correct. Similar to the 'k' equation, the transport equation for 'ε' can now be given as

$$\frac{\partial(\rho\varepsilon)}{\partial t} + \frac{\partial(\rho\varepsilon U_i)}{\partial x_i} = \frac{\partial}{\partial x_i}\left[\left(\mu + \frac{\mu_t}{\sigma_\varepsilon}\right)\frac{\partial\varepsilon}{\partial x_i}\right] + C_{1\varepsilon}\frac{\varepsilon}{k}2\mu_t E_{ij} \cdot E_{ij} - C_{2\varepsilon}\rho\frac{\varepsilon^2}{k} \qquad (5.25)$$

In practice, the rate at which the large eddies extract energy from the mean flow is precisely matched to the rate of transfer of energy across the energy spectrum to small, dissipating eddies. If this was not the case the energy at some scales of turbulence could grow or diminish without limit. Therefore, it is justified to express the large scale turbulent parameters, velocity 'ϑ' and length 'ℓ' scale, in terms of small scales 'k' and 'ε' as follows

$$\vartheta = k^{\frac{1}{2}}, \ \ell = \frac{k^{\frac{3}{2}}}{\varepsilon} \qquad (5.26)$$

The turbulent/eddy viscosity can be specified by applying the same approach as in the mixing length model which is as follows

$$\mu_t = C\rho\vartheta\ell = \rho C_\mu\frac{k^2}{\varepsilon} \qquad (5.27)$$

The equations (5.24), (5.25) and (5.27) contain five adjustable constants. The standard '$k-\varepsilon$' model employs values for the constants that are arrived at by comprehensive data fitting for a wide range of turbulent flows

$$C_\mu = 0.09, \ \sigma_k = 1.00, \ \sigma_\varepsilon = 1.30 \ C_{1\varepsilon} = 1.44 \ \& \ C_{2\varepsilon} = 1.92$$

The standard '$k-\omega$' model proposed by Wilcox is based on transport equations for the turbulent kinetic energy 'k' and the turbulent frequency of the eddies 'ω'. The turbulence frequency has the dimensions of '$\frac{1}{time}$' and is often called the specific dissipation rate because it can also be thought of as the ratio of 'ε' to 'k' ($\omega = \frac{\varepsilon}{C_\mu k}$). Writing the turbulence large scales in terms of 'k' and 'ω' as

$$\vartheta = k^{\frac{1}{2}}, \ \ell = \frac{k^{\frac{1}{2}}}{\omega} \tag{5.28}$$

The turbulent viscosity is now given by

$$\mu_t = C\rho\vartheta\ell = \rho\frac{k}{\omega} \tag{5.29}$$

The constant of proportionality 'C' is accounted for within the turbulent frequency 'ω'. The standard model uses the following transport equations for 'k' and 'ω'

$$\frac{\partial(\rho k)}{\partial t} + \frac{\partial(\rho k U_i)}{\partial x_i} = \frac{\partial}{\partial x_i}\left[\left(\mu + \frac{\mu_t}{\sigma_k}\right)\frac{\partial k}{\partial x_i}\right] + 2\mu_t E_{ij} \cdot E_{ij} - \beta^*\rho\omega k \tag{5.30}$$

$$\frac{\partial(\rho\omega)}{\partial t} + \frac{\partial(\rho\omega U_i)}{\partial x_i} = \frac{\partial}{\partial x_i}\left[\left(\mu + \frac{\mu_t}{\sigma_\omega}\right)\frac{\partial\omega}{\partial x_i}\right] + \alpha\frac{\omega}{k}2\mu_t E_{ij} \cdot E_{ij} - \beta\rho\omega^2 \tag{5.31}$$

The model constants are

$$\beta^* = 0.09 = C_\mu, \ \sigma_k = 2, \ \sigma_\omega = 2, \ \alpha = 0.555 \ \& \ \beta = 0.075$$

The turbulence models can now be adapted to multiphase flows in a similar manner as presented in section 5.1.1 where multiphase conservation equations are derived from the single phase conservation equations using the concept of the Two-Fluid Model. Consequently, the most general multiphase turbulence model derives the transport equations for each phase which are coupled by means of additional production '$P_{Interface}$' and destruction terms '$D_{Interface}$' at the gas-liquid interface. For the wavy stratified flows, this means that the turbulent transport of mass, momentum and energy in both phases at and in the vicinity of the gas-liquid interface should be thoroughly known because only then it is plausible to model the additional production '$P_{Interface}$' and destruction terms '$D_{Interface}$' precisely. As mentioned earlier in section 2.2, due to practical reasons, there is a lack of understanding at and in the vicinity of the gas-liquid interface and this is still an area of intense research. Therefore, stratified flows almost in all cases are simulated with a homogenous turbulence modelling approach, i.e. a single turbulence field is calculated for all phases, irrespective of the fact, if the Two-Fluid or the homogenous model is used to resolve the flow field. This concept is similar to the homogeneous approximation (see section 5.1.2) of the Two-Fluid model. Accordingly, the transport equations of the '$k-\varepsilon$' (Eqs. 5.24-5.25) & '$k-\omega$'

(Eqs. 5.30-5.31) turbulence models and relations for turbulent viscosities (Eq. 5.27 & 5.29), employ phase-averaged physical properties (Eqs. 5.14-5.15) to establish a coupling between the turbulence fields of the two phases. Based on the fact that the preferred multiphase model in this research work is a homogenous model (VOF) which can only accept a homogenous turbulence model, it can be said that the specifications of the continuum and turbulence modelling perfectly supports the approach of this thesis.

5.3   Application of the VOF Model to Thick Film Flows

To analyse the VOF (free surface/homogenous) model, a test case of practical importance where interface shear stress is the major driving force is simulated using two commercial CFD codes; ANSYS CFX 12.0 and FLUENT 6.3. The experimental data is taken from the well-known published work of Fabre et al. (1987) that is also used by several other authors e.g. Mouza et al. (2001) to validate their CFD methodologies. The data used here corresponds to an air-water horizontal rectangular channel, $0.2m$ wide, $0.1m$ high and $12m$ long, and simulates fully developed flow conditions. The flow in the test channel manifested two-dimensional behaviour after fully developed conditions were reached; therefore, the CFD test domain used here is also two-dimensional, $0.1 \times 1m$, and contains about $10,000$ cells. The number of cells used gives a $0.5mm$ resolution near the wall and the interface. In the preceding studies, the interface was resolved up to $0.1mm$ and even finer using an automatic mesh adaption scheme but no considerable advantages could be observed over a $0.5mm$ resolution. The setup includes '$k-\varepsilon$' turbulence model, $0.038m$ initial liquid film thickness and fully developed conditions. The simulation is transient to account for a changing interface topology (small 2D waves travelling on the film surface). An adaptive time step controlled by the maximum Courant number of 5.0 (RMS Courant number usually less than 1) is used to calculate the flow field. This is a very reasonable approach for shear driven flows when handled with the homogenous model (VOF) due to the completely different maximum velocities of the two phases.

In Fig. 5.2, the velocity profile along the height of the channel as calculated by the VOF method is compared with the experimental data. As evident from the figure, VOF predicts an unphysical gas side velocity profile. The velocity gradient near the interface is significantly underestimated and the maximum gas side velocity is shifted towards the top wall. In addition to the unphysical gas side velocity profile, VOF also greatly overpredicted the pressure losses ($Experimental\ \Delta P = 2.1Pa/m,\ Calculated\ \Delta P \cong 10Pa/m$). Other researchers (e.g. Frank (2005)) have also reported a similar trend in velocity profiles with high velocity gradients across the interface. This behaviour of VOF remained consistent with the other two equation turbulence model '$k-\omega$' and different Reynolds stress turbulence models which were also investigated. Grid refinement resulted in almost no improvement which is also experienced by other researchers e.g. Strakey (2004). Based on a qualitative observation, LES has shown some improvement (Strakey (2004)). However, later investigations e.g. Reboux et al. (2006), Liovic et al. (2007) have revealed that even LES needs some interface treatment before any significant, plausible quantitative validation could be reached.

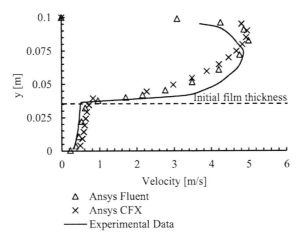

Fig. 5.2: Velocity distribution in both phases

5.3.1 Investigation of the Model Failure

To be able to isolate the exact cause of the VOF model's unphysical behaviour near the gas-liquid interface, the test case is further simplified with the introduction of laminar flow conditions. The 2D CFD test domain remains the same as before. For validation, the exact solution of the gas and liquid velocity profiles, Eqs. 5.32 - 5.33 (taken from Thome (2004)), is used.

$$v_L(y) = \frac{1}{2\mu_L}\frac{\mathrm{d}P}{\mathrm{d}x}\left[(\delta - H)^2 - \delta^2\right] + \frac{\tau_i}{\mu_L}y \ where \ 0 \le y \le \delta \tag{5.32}$$

$$v_G(y) = \frac{1}{2\mu_G}\frac{\mathrm{d}P}{\mathrm{d}x}\left[(y - \delta)^2 - (H - \delta)^2\right] + \frac{\tau_i}{\mu_G}(y - H) \ where \ \delta \le y \le H \tag{5.33}$$

In Fig. 5.3, the velocity profile predicted by the VOF method and the exact solution are plotted. As can be seen from the figure, an excellent agreement is found between the VOF prediction and the exact solution. Since an accurate velocity profile leads to a correct shear stress distribution, hence pressure drop, no further comparisons are shown here to prove VOF accuracy for the laminar case. This part of the investigation strengthens the anticipation that the shortcoming in the VOF model is coming from an inadequate turbulence modelling. According to the governing physics at the gas-liquid interface, a gas boundary layer establishes on the interface where viscous stresses take over from the turbulent Reynolds stresses. The standard '$k-\varepsilon$' turbulence model is only designed for high Reynolds numbers (core flow) and is unable to reproduce the near wall (interface) behaviour. The flow behaviour near the wall is modelled either by making use of the wall function approach or with the help of low Reynolds number turbulence models. To understand the behaviour of the homogenous turbulence modelling approach near the gas-liquid interface, the turbulent quantities profiles corresponding to the velocity profile shown in Fig. 5.2 are plotted in Figs. 5.4-5.6.

Fig. 5.4 compares the turbulent kinetic energy distribution along the height of the channel with the qualitative wall behaviour near the interface (the dashed line represents the initial location of

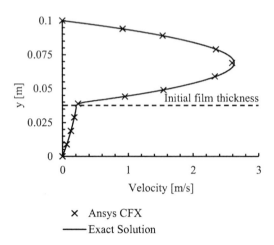

Fig. 5.3: Laminar velocity distribution in phases

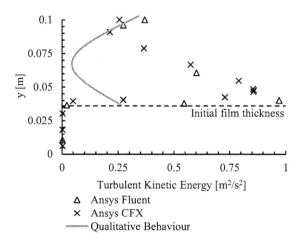

Fig. 5.4: Turbulent kinetic energy distribution in phases

the interface). From the rough wall analogy, it can be said that the turbulence model predicts a very high turbulent kinetic energy production near the interface. To explain the unphysical high production of turbulent kinetic energy, the term (Eq. 5.34) in the transport equation (Eq. 5.24) responsible for the production of turbulent kinetic energy is analysed in the next paragraph.

$$P^{(k)} = 2\mu_t E_{ij} \cdot E_{ij} \qquad (5.34)$$

The production term in the turbulence models is based on the physical reasoning that the highest values of the Reynolds stresses '$-\rho \overline{u_i' u_j'}$' are found in the region where the mean velocity gradient is highest, highlighting the intimate connection between the turbulence production and the sheared mean flow (Versteeg and Malalasekera (1995)). A very important point to be noted here is that the homogenous turbulence model employs the same transport equations as for the single phase turbulent flows and there is no additional term which models the gas-liquid interface physics. This means that a single turbulence field is calculated for both phases without taking into account the governing turbulence physics near the gas-liquid interface. On the other hand, in the turbulence transport equations, the two completely different phases are coupled with the help of volume fraction averaged physical properties (Eqs. 5.14-5.15). The large difference in physical properties across the interface will also result in some unphysical properties within the interface area due to the averaging procedure. For shear driven liquid film flows, large velocity and physical properties gradients exist across the interface. Due to homogenous turbulence modelling, this will result in an unphysical high turbulent kinetic energy production at the gas side near the interface which will be transported farther in the gas domain as can be seen in Fig. 5.4.

In Fig. 5.5, the turbulent eddy dissipation is plotted against the height of the channel. From the rough wall analogy, it can be seen that the dissipation rate is much lower at the interface than expected (qualitative wall behaviour). To explain the low dissipation, the source term (Eq. 5.35)

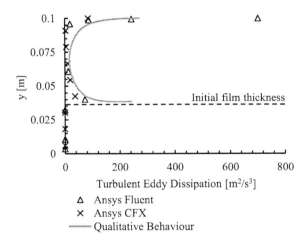

Fig. 5.5: Turbulent eddy dissipation distribution in phases

in the transport equation for 'ε' (Eq. 5.25) in '$k-\varepsilon$' turbulence model is analysed in the next paragraph.

$$P^{(\varepsilon)} = C_{1\varepsilon}\frac{\varepsilon}{k}2\mu_t E_{ij} \cdot E_{ij} \qquad (5.35)$$

Near the wall, the measurements of turbulent kinetic energy budgets indicate that the rate of turbulence production equals the rate of dissipation i.e. the regions of highest turbulent kinetic energy production are also the regions of highest turbulent dissipation production (Versteeg and Malalasekera (1995)). This is why, the production and destruction terms in the dissipation transport equation are assumed to be proportional to the production and destruction terms of the turbulent kinetic energy transport equation (see section 5.2.2). In the vicinity of the gas-liquid interface, the viscous dominated region requires that the viscous stresses take over from the turbulent stresses. However, in the standard turbulence model, there is no term in the dissipation transport equation which can provide this necessary transition from the turbulent to the viscous stresses. Therefore, the low dissipation near the interface can be explained on the basis of the lack of appropriate turbulence modelling for the two-phase wavy stratified flows. With reference to the dissipation transport equation (Eq. 5.25), it can now be argued that the dissipation source term (Eq. 5.35) is insensitive to the viscous effects present near the interface. Analogous to the low Reynolds number turbulence models (Patel et al. (1985)), additional damping functions are necessary to include the viscous contribution in the production and destruction terms of the dissipation equation. As a consequence, the turbulent kinetic energy produced at the gas-liquid interface cannot be destroyed at or in the vicinity of the interface. On the other hand, as learned earlier in the turbulence modelling section, this turbulent kinetic energy has to be destroyed somewhere in the domain. As can be seen from Fig. 5.5, this happens on the top wall where unusually high dissipation is evident; consequently, there is a shift of maximum velocity near the top wall (see Fig. 5.2).

The turbulent viscosity is derived from the turbulent kinetic energy and its dissipation rate. More precisely, the turbulent viscosity is proportional to the turbulent kinetic energy and is inversely proportional to the turbulent eddy dissipation (Eq. 5.27). From the definition, it is clear that the uncontrolled production of turbulent kinetic energy near the interface will lead to unphysical high turbulent eddy viscosity in the shear driven flows near the interface. In Fig. 5.6, the turbulent eddy viscosity distribution in the gas phase is plotted. From the rough wall analogy (qualitative wall behaviour), it can be seen that the turbulent eddy viscosity near the interface is still very high where (physically) it should decay to zero due to the action of viscosity in the gas side boundary layer. The unphysical high turbulent eddy viscosity enters the momentum equation through the Reynolds's stress term '$-\rho \overline{u_i' u_j'}$' and causes an additional drag; therefore, lowering the gas velocity below the experimental observation. This conclusion can also be drawn by looking once again at the near interface gas side velocity profile (Fig. 5.2) which resembles laminar although the flow is turbulent. This evidence can now be used to explain an additional momentum transfer (false momentum transfer) between the phases realized through turbulent stresses.

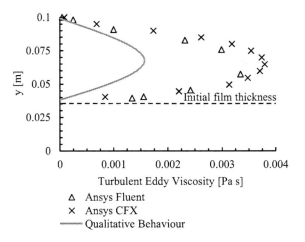

Fig. 5.6: Turbulent viscosity distribution in the gas phase

5.3.2 Turbulence Model Behaviour near the Interface

Based on the investigations carried out in this section, inadequate turbulence modelling can be identified as the major cause governing the unphysical model behaviour near the gas-liquid interface. The gas-liquid interface is traced with the help of an additional transport equation (Eq. 5.12) but has no physical meaning in the flow governing equations. This means that the interface is effectively non-existent because there are no interface specific boundary conditions implemented either in the VOF model or the turbulent transport equations. This lack of information on the interface allows for an additional momentum transfer to the liquid film which is realized through turbulent stresses. From the theory of classical turbulence modelling, this means that the eddies of each phase can move freely through the gas-liquid interface (see Fig. 5.7), which does not reflect the physics of shear driven liquid wall films. Keeping this in mind and accordingly allowing the eddies of the fast moving lighter phase to enter the slow moving heavier phase explains the additional momentum transfer to the liquid film (overpredicted velocity profile, see Fig. 5.2). In a similar manner, by allowing the eddies of the slow moving heavier phase to enter the fast moving lighter phase, the additional drag on the gas phase can be explained. The same phenomenon can also be used to explain the shift of maximum gas velocity towards the top wall and high pressure losses which are manifested in the predicted pressure losses.

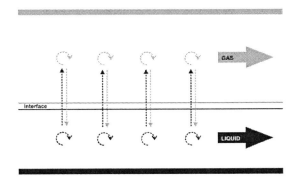

Fig. 5.7: Unphysical eddying motion across the interface due to inadequate turbulence modelling

5.4 Improved VOF Model

In the last section, it is shown that due to inadequate turbulence modelling at the interface the VOF model cannot handle turbulent shear driven film flows. In this section, a mathematically closed interface treatment is introduced which accounts for the necessary boundary condition at the interface.

5.4.1 Previous Approaches

Before going into the details of the proposed mathematical formulation, the open literature on the solution strategies (see section 2.4 for details) is briefly revised to establish the absolute necessity and superior characteristics of the interface treatment technique presented in this work. In literature, two approaches can be found which are discussed in the following:

5.4.1.1 Moving Rough Wall Analogy

This approach makes use of a moving rough wall analogy e.g. Mouza et al. (2001), Ghoria et al. (2005). Correspondingly, a wall function is used near the interface. For this approach to work, the interface coordinates should be known in advance where an additional switching variable similar to 'y^+' (wall coordinates) is defined. Waves appearing on the interface are seen as roughness which can be experimentally determined. Ghorai et al. (2005) used a similar approach to successfully improve the VOF method for the shear driven films and points out the other necessary (experimental driven) information which is as follows:

- Classical turbulent velocity distribution law with experimentally determined roughness,

- Interface velocity and

- Turbulent kinetic energy as well as its dissipation rate near the interface

One drawback of this approach is its limited applicability i.e. it only works well for the flow conditions for which it is experimentally determined. Another problem is, for undeveloped flows the surface roughness is variable; however, in this approach a constant equivalent surface roughness value is used. Hence, if undeveloped flows are of interest, this approach will not behave correctly. Besides this, as explicitly mentioned by Mouza et al. (2001) in his work, wave generation, evolution and propagation and the resulting secondary flows in the gas phase cannot be modelled using this approach. Therefore, if the local flow field features (free surface dynamics, secondary flows etc...) cannot be neglected, this approach will not lead to plausible results. Moreover, the biggest disadvantage of this approach lies in the fact that the interface shear stress has to be provided as an input to the CFD calculations which is normally an unknown for all cases of practical importance and desired as an output.

5.4.1.2 Adaptation of the Turbulence Modelling

This approach adapts the homogenous turbulence model to the wavy stratified flows such that the right behaviour of the turbulence transport mechanism results automatically in both phases as well as near the interface. This also adjusts the gas side velocity profile resulting in the development of a turbulent boundary layer on top of the free surface where viscous stresses take over from the turbulent stresses. Consequently, the false momentum transfer across the gas-liquid interface is eliminated and the wall film flows under the influence of only viscous stresses which is the governing physics of the shear driven wall films. Starting from the early 1980's (Akai et al. (1980,

1981), see section 2.2.2 for details), many researchers have contributed to this approach and performed successful validation of their results with the experimental data. However, analogous to the first approach, this approach also showed a strong case dependent behaviour when employed under circumstances different from those for which it was first formulated. The problem lies primarily in the turbulence modelling adaptation which requires a detailed knowledge on the behaviour of turbulent transport quantities in the vicinity of the gas-liquid interface. Several researchers who have contributed to this approach have also attempted to determine the behaviour of turbulent quantities experimentally and formulated suitable interfacial boundary conditions. However, due to the limitations on instrumentation, the behaviour of turbulent quantities could only be measured up to the near interface but not up to the interface especially in the presence of waves travelling on the free surface. Therefore, indicating a lack of information at the location where it is most needed. Hence, explaining the lack of generality in this approach. Realizing this fact, other researchers e.g. Issa (1988), Egorov (2004) have used physical reasoning to formulate the necessary interfacial boundary conditions for the turbulent quantities but till today no generalized boundary conditions could be presented. Nevertheless, such an approach will deliver interface shear stress as an output instead of being required as an input as is the case with the first approach and is also capable of resolving the free surface dynamics (wave generation, evolution and propagation). This feature makes this approach very attractive for the practical engineering purposes. Therefore, within the present work, this approach is further developed to formulate a generalized boundary condition. In this thesis onwards, this approach is referred to as the "interface treatment". Whenever results are compared, the original VOF method is referred to as the case without interface treatment.

5.4.2   Interface Treatment Method

To derive the exact boundary condition for the turbulent quantities, the exact source of problem in the turbulence modelling and the physics of turbulence near the interface must be known in advance. The concept of false momentum transfer introduced at the end of section 5.3, identifies the exact source of problem in the turbulence modelling and can be summarized in the following points:

- The homogenous turbulence modelling approach allows the exchange of turbulent momentum (false momentum transfer) across the interface i.e. an eddy from the high velocity gas phase can enter the liquid domain and vice versa.

- This results in a high turbulent kinetic energy production on the gas side, near the gas-liquid interface due to high velocity and physical properties gradients.

- The turbulent viscosity in the two-equation turbulence models (eddy viscosity turbulence models) is proportional to the turbulent kinetic energy and the averaged material densities. Hence, an unphysically high turbulent viscosity results on the gas side near the gas-liquid interface.

- The additional turbulent kinetic energy is also transported farther in the gas domain resulting in an unphysical gas velocity field near the gas-liquid interface.

- The result is a relatively small velocity gradient in the gas domain near the gas-liquid interface and a profile rather laminar than of turbulent nature.

- Another consequence of a global importance is an unphysically high pressure drop in the gas phase for a velocity/mass flow inlet boundary condition or an unphysical small gas discharge for a total pressure inlet boundary condition.

- In the liquid film, unphysically high turbulent momentum is transferred from the gas phase and results in an overpredicted liquid velocity.

On the other hand, the physics of turbulence and the expected behaviour of a turbulence model near the gas-liquid interface is as follows:

- From theory, interface behaves like a moving wall and the waves appearing on an interface act as surface roughness.

- Due to a velocity gradient across the interface, a turbulent boundary layer develops in the gas phase above the gas-liquid interface.

- Accordingly, the no-slip condition at the free surface implies that adjacent to the free surface a laminar sublayer exists. Therefore, in this region viscous stresses dominate and all turbulent stresses decrease sharply to zero.

- The turbulent viscosity in this region, therefore, should also tend to zero.

- This indicates that the production of turbulent kinetic energy needs to be damped in the gas phase near the gas-liquid interface.

- The criteria that the turbulent viscosity near the interface must be negligibly small, gives the additional amount of dissipation of the turbulent kinetic energy.

- In this manner, the false momentum transfer through turbulent stresses across the interface can be restricted to a negligible value.

- The gas velocity gradient near the interface also assumes its physical distribution and the momentum transfer then results by the shear (viscous) stress at the interface.

In Fig. 5.8 the closed loop interaction between the turbulent quantities and the momentum equation is shown. In words, the turbulent quantities close the averaged momentum equations by providing the turbulent stresses. The velocity field is calculated and the turbulent quantities are updated based on the newly calculated velocity field. This procedure is repeated until a converged solution is reached.

The turbulence physics near the free surface suggests that the turbulent viscosity, analogous to a wall, should be modelled in a way to get a minimum near the free surface. As can be seen from Fig. 5.8, this can be achieved by modifying any of the parameters playing a role in the turbulence production near the interface. However, care should be taken as modifying these quantities may result in abrupt discontinuities in the flow domain; hence, problems in convergence or an immediate solver breakdown. Considering these limitations, two improvement possibilities which have proved to be very robust for engineering relevant problems are introduced in this thesis.

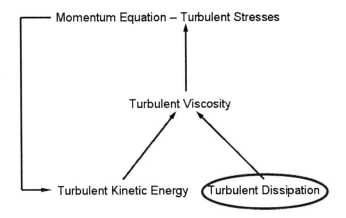

Fig. 5.8: Calculation of turbulent stresses in the momentum equation

5.4.2.1 Simplified Interface Treatment

The simplified interface treatment makes use of the smeared interfaces, as the interface is not sharp but distributed across a number of cells (usually three in commercial CFD codes e.g. ANSYS CFX and FLUENT), and directly adjusts the turbulent viscosity there according to the rule shown below:

$$\mu_t = \left|\begin{array}{l} \to 0 \; if \; 0 < Volume \; fraction \; of \; gas < 1 \\[2mm] \rho C_\mu \frac{k^2}{\varepsilon} \; , \; \rho \frac{k}{\omega} \; otherwise \end{array}\right. \tag{5.36}$$

This approach is very simple to implement and works fairly well. A closer look at this approach indicates that this approach works by simply setting the turbulent viscosity in the production and gradient diffusion terms of the turbulent kinetic energy transport equation (see Eq. 5.24) to a very small value. However, the molecular viscosity of the heavier phase is usually not negligible; therefore, the diffusion of the turbulent kinetic energy from the heavier to the lighter phase cannot be completely avoided. Hence, the false momentum transfer between the phases cannot be fully eliminated in this approach. Due to this reason, as will be shown in the following section, a further optimization potential in the turbulence modelling remains and can be utilized.

5.4.2.2   Refined (Mathematically Closed) Interface Treatment

A more refined approach towards the interface treatment can be developed by indirectly altering the turbulent viscosity to the desired value in the vicinity of the gas-liquid interface. This can be done either by a sink term (additional destruction term) in the turbulent kinetic energy transport equation or by a source term (additional production) in the dissipation 'ε'/specific dissipation 'ω' transport equation (see Fig. 5.8). Since the destruction term in the turbulent kinetic energy transport equation is given by the dissipation transport equation, it is natural to use the dissipation transport equation for this purpose. Therefore, in this work, the correct behaviour of turbulent viscosity near the interface is achieved by a local increase in the dissipation production i.e. by adding a source term in the dissipation/specific dissipation transport equation. However, the difficulty now is how to determine the right amount of additional dissipation and its location because if its value is lower than required, the method will not improve the flow field. On the other hand, more than required dissipation can result in an unphysical turbulent kinetic energy ($k \leq 0$) which may also be numerically unstable because in the 'ε'/'ω' transport equation (Eq. 5.25/5.30), 'k' appears in the denominator; hence, a zero value will be undefined and a negative value will change the physical interpretation of the terms by changing the sign. Therefore, similar to the previous works of many researchers (see section 2.2.2), the questions now arise what should be the right magnitude of this additional dissipation/specific dissipation production and where should it be added to achieve the right behaviour of the turbulent flow field. The quantification procedure of the source term resembles the previous works of few researchers e.g. Issa (1988), Egorov (2004) only to the extent that a physical reasoning is used to make the basic assumptions. In contrary to the previous works, the quantification procedure here is completely unique in the sense that a mathematically closed relationship is derived from the two-equation turbulence models and no other input is necessary. This characteristic of the introduced interface treatment technique makes it absolutely exceptional as compared to all the previous works in the last three decades. The right magnitude of the additional dissipation can now be determined by the turbulent quantities themselves as shown in the following:

The turbulent viscosity at a point near the free surface as calculated by the solver is as follows

$$\mu_t = \rho C_\mu \frac{k^2}{\varepsilon} \, , \, \mu_t = \rho \frac{k}{\omega} \tag{5.37}$$

Rearranging Eq. 5.37 for dissipation/specific dissipation, yields the exact value of the dissipation/specific dissipation required to destroy the complete (unphysical) turbulent kinetic energy produced near the free surface

$$\Rightarrow \varepsilon = \frac{\rho C_\mu}{\mu_t} k^2 \, , \, \omega = \frac{\rho}{\mu_t} k \tag{5.38}$$

The right magnitude of the dissipation/specific dissipation is now a known value but needs a further elaboration before it can be added as a source term in the dissipation/specific dissipation transport equation. Therefore, for a dimensionally correct source term, Eq. 5.38 is differentiated with respect to time to get the dissipation rate. This results in a dissipation rate (Eq. 5.39) in terms of the rate of change of turbulent kinetic energy. *(Note: The source term is derived on per time step basis and is also calculated and added every time step. Therefore, density can be treated as a constant during differentiation with respect to time)*

$$\Rightarrow \frac{\partial \varepsilon}{\partial t} = \frac{2kC_\mu}{\mu_t} \frac{\partial(\rho k)}{\partial t} \;,\; \frac{\partial \omega}{\partial t} = \frac{1}{\mu_t} \frac{\partial(\rho k)}{\partial t} \tag{5.39}$$

The rate of change of turbulent kinetic energy in a control volume is given by the transport equations Eq. 5.24 $(k-\varepsilon)$ and Eq. 5.30 $(k-\omega)$. Near the interface, the adjacent flow is dominated by viscous forces (laminar sublayer) and the turbulent stresses are zero. Therefore, the rate of production of turbulent kinetic energy in this region and its transport across the interface should be zero. Keeping this in mind, the transport equation for the rate of change of turbulent kinetic energy can be simplified by neglecting the transport terms. By realizing the fact that the sink terms '$\rho\varepsilon$'/'$\beta^* \rho \omega k$' are already accommodated in the turbulent viscosity relations given by Eq. 5.37, they can also be neglected. Accordingly, near the interface, the turbulent kinetic energy rate in Eq. 5.39 can be replaced by the respective production terms of Eq. 5.24 and Eq. 5.30.

$$\Rightarrow \frac{\partial \varepsilon}{\partial t} = \frac{2kC_\mu}{\mu_t}(2\mu_t E_{ij} \cdot E_{ij}) = 4kC_\mu E_{ij} \cdot E_{ij} \;,\; \frac{\partial \omega}{\partial t} = \frac{1}{\mu_t}(2\mu_t E_{ij} \cdot E_{ij}) = 2E_{ij} \cdot E_{ij} \tag{5.40}$$

Eq. 5.40 expresses the dissipation rate per unit mass. The dissipation rate per unit volume is achieved by multiplying the equation with density

$$\Rightarrow \frac{\partial(\rho\varepsilon)}{\partial t} = 4k\rho C_\mu E_{ij} \cdot E_{ij} \;,\; \frac{\partial(\rho\omega)}{\partial t} = 2\rho E_{ij} \cdot E_{ij} \tag{5.41}$$

The above equation is now dimensionally correct and can be added as a source term in the dissipation/specific dissipation transport equation. The tensor 'E_{ij}' is the mean rate of deformation and is given by (in matrix form):

$$E_{ij} = \frac{\partial U_i}{\partial x_i} = \begin{pmatrix} E_{11} & E_{12} & E_{13} \\ E_{21} & E_{22} & E_{23} \\ E_{31} & E_{32} & E_{33} \end{pmatrix} = \begin{pmatrix} \frac{\partial u}{\partial x} & \frac{\partial u}{\partial y} & \frac{\partial u}{\partial z} \\ \frac{\partial v}{\partial x} & \frac{\partial v}{\partial y} & \frac{\partial v}{\partial z} \\ \frac{\partial w}{\partial x} & \frac{\partial w}{\partial y} & \frac{\partial w}{\partial z} \end{pmatrix} \tag{5.42}$$

The product '$E_{ij} \cdot E_{ij}$' in Eq. 5.41 is a dot/scalar product of two symmetric tensors and can be expanded as follows

$$E_{ij} \cdot E_{ij} = E_{11}^2 + E_{22}^2 + E_{33}^2 + 2E_{12}^2 + 2E_{13}^2 + 2E_{23}^2 \tag{5.43}$$

Substituting the values of the individual tensor components from Eq. 5.42 in Eq. 5.43

$$\Rightarrow E_{ij} \cdot E_{ij} = \left(\frac{\partial u}{\partial x}\right)^2 + \left(\frac{\partial v}{\partial y}\right)^2 + \left(\frac{\partial w}{\partial z}\right)^2 + \frac{1}{2}\left(\frac{\partial u}{\partial y} + \frac{\partial v}{\partial x}\right)^2 + \frac{1}{2}\left(\frac{\partial u}{\partial z} + \frac{\partial w}{\partial x}\right)^2 + \frac{1}{2}\left(\frac{\partial v}{\partial z} + \frac{\partial w}{\partial y}\right)^2 \tag{5.44}$$

Eq. 5.44 represents the full three dimensional mean flow gradients required to calculate the production term of the turbulence transport equations. For a two dimensional case, Eq. 5.44 can be simplified to

$$\Rightarrow E_{ij} \cdot E_{ij} = \left(\frac{\partial u}{\partial x}\right)^2 + \left(\frac{\partial v}{\partial y}\right)^2 + \frac{1}{2}\left(\frac{\partial u}{\partial y} + \frac{\partial v}{\partial x}\right)^2 \tag{5.45}$$

For a simple shear flow e.g. in boundary layers, there is only one considerable velocity gradient and the rest are negligible. If 'u' is the flow velocity in the flow direction 'x', and 'y' is the space-coordinate perpendicular to the flow direction, then '$E_{ij} \cdot E_{ij}$' can be further simplified as

$$\Rightarrow E_{ij} \cdot E_{ij} = \frac{1}{2} \left(\frac{\partial u}{\partial y} \right)^2 \tag{5.46}$$

Equations 5.44-5.46 give the last unknown in the source term (Eq. 5.41) for the dissipation/specific dissipation transport equation. The required gradients of the mean flow velocities are usually directly available from a solver.

5.4.2.3   Location of the Interface Treatment Source Term

Now to answer the question where the additional dissipation should be added, consider the following. Analogous to the low Reynolds number turbulence models, damping needs to be applied near the (wall like) interface in the gas phase to ensure that at low Reynolds numbers the viscous stresses take over from the turbulent stresses in the viscous dominated sublayer adjacent to the free surface. The location of the source term is, therefore, adjacent to the free surface in the gas domain. To ensure that no turbulent kinetic energy is either produced at or transported across the interface, it is necessary to add the additional dissipation in the cells containing the highest gradient in the vicinity of the interface. Quantitatively, it can be said that usually a distance equals to three times the interface resolution above the free surface, as shown in Fig. 5.9, works very well. For smooth interfaces and the cases where the film thickness is known, this can easily be

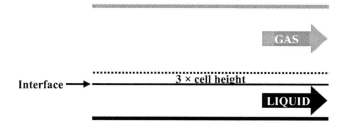

Fig. 5.9: Location of the source term above the interface

done by selecting the appropriate height coordinate. However, in most practical situations, the film thickness is an unknown and time dependent. To solve this problem, this work employs an independent concept to locate the appropriate cells above the interface. This concept states that the interfacial transfer of mass, momentum and energy are directly dependent on the contact surface area between the phases. This contact surface area is characterized by the interfacial area per unit volume between the phases, known as the Interfacial Area Density 'A', and has the dimensions of one over length '$(\frac{1}{m})$'. For a system of two continuous phases ($k = G, L$), Interfacial Area Density can be calculated in the following manner:

$$A_{GL} = |\nabla \alpha_G| \tag{5.47}$$

In commercial codes (e.g. ANSYS CFX), 'A_{GL}' usually marks three cells on either side of the interface where the interface can be assumed to locate at $\alpha \approx 0.5$. The Interfacial Area Density 'A' has usually values ranging around $\sim 1000 m^{-1}$ in the cells occupied by the interface and $1 m^{-1}$ elsewhere. Now to locate the cells above the interface, the following mathematical expression is proposed.

$$\lambda = \underbrace{\left(\frac{A-1}{A+1}\right)}_{I} \underbrace{\alpha_G}_{II} \tag{5.48}$$

where 'λ' is a switching function which activates the additional source term in the dissipation/specific dissipation transport equation (Eq. 5.30 & Eq. 5.36). Term I ensures that 'λ' is close to one in the cells occupied by the interface and zero elsewhere. Term II ensures that 'λ' is only active in the part of interface in the gas domain and returns to zero in the liquid domain. The modified transport equations can now be rewritten as follows.

$$\frac{\partial(\rho\varepsilon)}{\partial t} + \frac{\partial(\rho\varepsilon U_i)}{\partial x_i} = \frac{\partial}{\partial x_i}\left[\left(\mu + \frac{\mu_t}{\sigma_\varepsilon}\right)\frac{\partial\varepsilon}{\partial x_i}\right] + C_{1\varepsilon}\frac{\varepsilon}{k}2\mu_t E_{ij} \cdot E_{ij} - C_{2\varepsilon}\rho\frac{\varepsilon^2}{k} + \underbrace{4\lambda k\rho C_\mu E_{ij} \cdot E_{ij}}_{source\ term}$$

$$\tag{5.49}$$

$$\frac{\partial(\rho\omega)}{\partial t} + \frac{\partial(\rho\omega U_i)}{\partial x_i} = \frac{\partial}{\partial x_i}\left[\left(\mu + \frac{\mu_t}{\sigma_\omega}\right)\frac{\partial\omega}{\partial x_i}\right] + \alpha\frac{\omega}{k}2\mu_t E_{ij} \cdot E_{ij} - \beta\rho\omega^2 + \underbrace{2\lambda\rho E_{ij} \cdot E_{ij}}_{source\ term} \tag{5.50}$$

$$where\ \lambda = \begin{vmatrix} 1\ above\ the\ free\ surface\ up\ to\ (see\ Fig.5.9) \\ 0\ elsewhere \end{vmatrix}$$

The source term can now be implemented in any commercial CFD code. In this research work, the source term is implemented in ANSYS CFX and a number of test cases are presented in the next chapter. The next chapter also shows the simple implementation of 'λ' in ANSYS CFX.

6 Simulation Results and Discussion

In this chapter, the interface treatment techniques derived in the last chapter are tested against several test cases. The first test case makes use of a 2D stratified channel and is improved using the simplified (Eq. 5.36) as well as the refined (Eq. 5.41) interface treatment techniques. It is shown that although the simplified interface treatment improves the flow field considerably; however, it still lacks the refined interface treatment quality. The second test case deals with the 3D undeveloped film flow investigations using the ITS stratified flow test rig and simulates seven test conditions. These include the analysis of the effect of shearing gas flow, effect of gravity and the influence of film physical properties. Only the refined interface treatment approach is used to improve the turbulence behaviour near the gas-liquid interface. The parallels are then drawn between the results with and without interface treatment and critically analysed for the importance and effectiveness of the interface treatment approach.

6.1 Test Case #1 - High Viscosity Liquid

The test case uses standard air for the light phase and a very high viscosity imaginary fluid for the heavy phase. The density of the heavier phase is also increased to increase the film's inertia. There are two advantages of using such an imaginary fluid for the heavy phase. First, the heavy phase will behave like a wall with theoretically no motion at all thus simplifying the flow physics and allowing for a better understanding. Second, the reference (validation) data including the distribution of turbulent quantities across the interface can easily be achieved by carrying out a single phase CFD calculation for the air domain bounded by the standard wall type boundary conditions only.

CFD Domain
The 2D CFD domain is $100 \times 3000mm$ and contains about 59,346 cells. The setup includes the standard '$k-\varepsilon$' turbulence model and a $10mm$ high viscous fluid acting as a wall. This is a steady state ANSYS CFX 12.0 simulation.

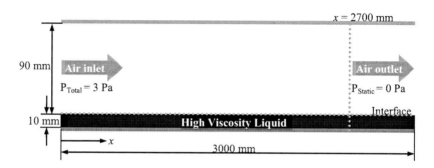

Fig. 6.1: CFD domain

Boundary Conditions
A total pressure inlet and a static pressure outlet are defined as boundary conditions for the required air flow. There is no inlet or outlet for the artificially made high viscosity fluid; instead, the fluid is confined between walls. It is safe to do so because the momentum transferred from the air is very small as compared to the very high inertia of the film and a very high dissipation in the fluid; hence, no significant movement is possible. The simplified as well the refined interface treatment (Eq. 5.41) together with the 2D expansion of the product of the mean rate of deformation (Eq. 5.45) are used to improve the turbulence behaviour across the interface. The height coordinate 'y' is used to implement the interface treatment in the cells as follows

$$\lambda = \begin{vmatrix} 1 \; if \; (y \leq 13mm) \\ 0 \; otherwise \end{vmatrix} \tag{6.1}$$

The switching term as given by Eq. 6.1 is active up to $3mm$ above the interface including the cells where the high viscosity fluid is present; hence, destroying any turbulent kinetic energy which might be produced in this region giving the desired and expected behaviour of the flow field to this simulation.

Results and Discussion
In Fig. 6.2a, the distribution of turbulent flow field quantities on a line $2700mm$ from the inlet are presented. The figure includes the reference data along with the data achieved with and without the interface treatment techniques. As can be seen from the figures, the VOF method in its original form largely deviates from the reference data. This deviation, as explained in the last chapter, comes from the false momentum transport across the interface and averaging of material properties in the interface area. It can also be seen from Fig. 6.2a that the simplified interface treatment approach significantly improves the behaviour of the turbulence quantities but not as good as the refined interface treatment approach.

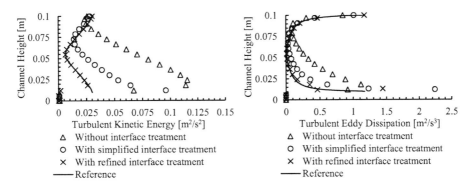

Fig. 6.2a: Turbulent kinetic energy distribution (Left); Turbulent eddy dissipation distribution (Right)

As stated in section 5.4.2.1, the reason why simplified interface treatment can only realize a limited improvement in the turbulence model behaviour lies in the underlying mechanism by which it works. Namely, the diffusion of the turbulent kinetic energy from the heavier to the lighter phase cannot be completely avoided; hence, the false momentum transfer between the phases cannot be completely avoided either. Fig. 6.2a (Left) also indicates this fact where as compared to the reference data, relatively high turbulent kinetic energy is evident at and above (in the gas domain) the interface. As a result, large values of turbulent viscosity are calculated above the interface as can be seen in Fig. 6.2b (Left). On the other hand, the refined interface treatment can also account for the diffusion of turbulent kinetic energy and therefore, for this test case, can reproduce an almost perfect behaviour of the turbulent quantities across the interface. Accordingly, as shown in Fig. 6.2b (Right), the velocity field calculated with the simplified interface treatment lacks the quality of refined interface treatment.

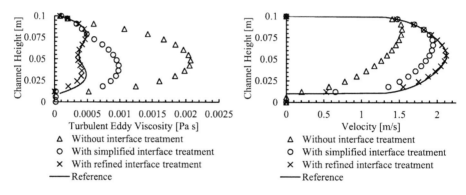

Fig. 6.2b: Turbulent viscosity distribution (Left); Velocity distribution (Right)

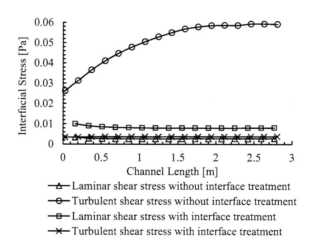

Fig. 6.3: Interface shear stress along the channel length

The effectiveness of the source term derived in the refined interface treatment approach can also be qualitatively judged by plotting the turbulent and laminar shear stresses at the interface along the channel length (see Fig. 6.3). A very high turbulent and a very low laminar shear stress are calculated by the VOF method (without interface treatment). This is not the governing physics near the interface which acts like a wall, hence, dampens the turbulent fluctuations. In contrary, the proportion of turbulent and laminar shear stresses with interface treatment corresponds to the physics near the free surface.

6.2   Test Case #2 - ITS Stratified Flow Test Rig

CFD Domain
A 1:1 CFD domain of the test rig is used for the numerical simulations. A structured mesh comprising of 2,661,624 elements is prepared in ANSYS ICEM. Turbulence is modelled using the '$k-\omega$' turbulence model. For correcting the turbulence behaviour across the interface, the dissipation source term given by Eq. 5.41 is employed. The dot product of the mean rate of deformations in Eq. 5.41 is given by the full 3D expansion (Eq. 5.44). Implementation of the source term in the desired cells is realized according to Eq. 5.48. The simulation is transient in order to correctly account for disturbances (waves) appearing on the film surface.

Boundary Conditions

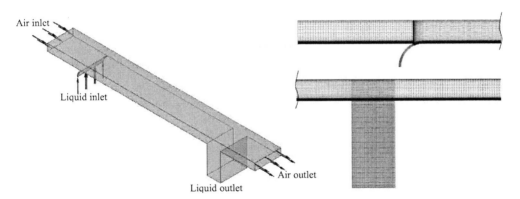

Fig. 6.4: CFD domain of the ITS stratified flow test rig

The air and liquid entries into the CFD domain are realized with the help of mass flow inlets. Mass flow is a robust inlet boundary condition for the solver and corresponds to the real conditions governing the air and liquid inlets of the test rig (see section 4.1.1). The film mass flow rate is kept constant at $0.138583 kg/s (\sim 8.34 l/min)$ for all tested conditions.

Air exits the CFD domain with the help of a pressure outlet which is given a gauge pressure as measured during the experiments. Liquid is collected in the separator tank and removed at a rate so as to maintain a constant liquid level. This is achieved by assigning the liquid outlet a wall type boundary condition with a conditional sink term. The implementation of this source term in ANSYS CFX pre-processor is as follows:

$$if(areaAve(pressure)@Liquid\ Outlet\ >\ 551 Pa, -0.145 kg/s, -0.125 kg/s) \quad (6.2)$$

The first term in Eq. 6.2 is a condition and checks if the area-averaged static pressure, which is a direct indication of the liquid level in the separator tank, is greater than a certain value. If the condition is true, the sink is assigned the 2nd term whereas if the condition is evaluated to be false the sink is assigned the 3rd term of Eq. 6.2. The mass flow given by the 2nd term is greater than the mass flow entering the separator tank and the mass flow given by the 3rd term is less than the mass flow entering the separator tank. In this manner, the expression works by extracting more liquid than entering if the liquid level is on the rise in the separator tank and keeps on extracting until the liquid level falls below the level of $551Pa$. From this point onwards, the sink extracts less liquid than entering, hence, allowing for the liquid level to rise again. This mechanism ensures an almost constant liquid level in the separator tank. All other surfaces are assigned a wall boundary condition.

Calculation of the Additional Momentum Transfer 'I' to Film

The additional momentum transferred to the film is calculated in a similar manner as in the experiments i.e.

$$I = \Delta P_{WF} - \Delta P_{WoF} \tag{6.3}$$

The CFX solver manager provides the flow of momentum (force) on the inlet and outlet surfaces. Based on this information, the total pressure drop in the channel can be calculated as follows.

$$\Delta P_{WF/WoF} = \frac{1}{A}\left(F_{Air\ Inlet} - F_{Air\ Outlet}\right) \tag{6.4}$$

The simulations are transient in nature. The film requires approximately three seconds to complete one pass of the test section. Therefore, all simulations are run for approx. four seconds and sometimes more to ensure that the flow regime (in terms of characteristic waveforms) does not change anymore. The results are then averaged over one second, usually from the 3rd to the 4th second.

Sources of Error

To keep the number of elements limited to a feasible range and the fact that resolving the tank flow phenomena is not the aim of this work, the gas-liquid interface in the separator tank is not fully resolved. This however, causes numerical diffusion; consequently, resulting in an unphysical fluid with averaged physical properties present after the test section i.e. in the separator tank '4' and the outlet module '5' (see Fig. 4.2). Hence, depending on the amount of unphysical fluid leaving the domain, the resistance increases especially if this unphysical fluid settles on the boundary walls because the force the CFX solver calculates on the inlet/outlet surfaces (Eq. 6.4) includes the viscous forces on the boundary walls. Therefore, for the cases with high numerical diffusion, the pressure drop with film in the test section is expected to be significantly underpredicted. Moreover, unlike the experiments where the average is conducted over at least $30secs$, the CFD results are averaged over only $1sec$. Therefore, this factor can also result in a discrepancy between the simulation results and the experimental data as can be seen e.g. in Tab. 6.1 where Mesh 2 shows an inconsistent result.

The surface disturbances/waves caused by a shearing gas flow on a film surface can cover a wide range of scales including the very small unresolvable scales. The numerical solution can only resolve the statistically averaged motion of the free surface including the waves which are not too small relative to the local mesh resolution. In case small unresolvable scales are anticipated to appear on the film surface and considered to be important in the overall flow field, the results can only be improved either by modelling (e.g. increased roughness) their influence on the averaged flow field or by refining the mesh locally.

Mesh Independency

The mesh is divided into two domains, namely, the gas domain which is primarily occupied by the gas phase and the film domain which is primarily occupied by the liquid film. To resolve a few millimetres thick film and to be able to resolve the gas-liquid interface with sufficient accuracy, the film domain cannot be meshed completely arbitrarily. Keeping in view the allowable mesh expansion factor and aspect ratio, the film thickness is resolved with $0.3mm$ high cells occupying the $5mm$ high (film) domain which is kept constant for all cases. The number of cells along the height of the gas domain and along the width of the channel is varied in two steps from a first relatively coarse mesh (Mesh 1) as shown in Table 6.1. This variation does not include the cells at walls which are kept at $0.5mm$ to give an average 'y^{+}' of 15 for the whole CFD domain. Based on the low numerical diffusion in the separator tank and the fact that a fair comparison can

Mesh	No. of Elements	Max. Element Size $(l \times b \times h)[mm]$		Additional Momemtum Transfer*$[Pa]$	Liquid Mass Flow**$(\%)$
Mesh1	677,484	gas	$3.0 \times 6.0 \times 7.3$	3.11	6.6
		liquid	$3.0 \times 6.0 \times 0.3$		
Mesh2	1,195,560	gas	$3.0 \times 3.6 \times 5.8$	2.83	6.3
		liquid	$3.0 \times 3.6 \times 0.3$		
Mesh3	2,661,624	gas	$3.0 \times 2.5 \times 3.0$	3.70	3.7
		liquid	$3.0 \times 2.5 \times 0.3$		

* Simulations include interface treatment

** Liquid flow rate at the air outlet due to numerical diffusion in the separator tank

Table 6.1: Mesh Independency

be achieved with the experimental data (section 6.2.1), Mesh 3 is selected for all simulations. A further refinement almost doubles the number of elements present in Mesh 3. Keeping in mind the fact that every test case has to be run for at least $4secs$, further refinements will result in a very high demand on computational power and time. On the other hand, the aim of this work is to prove the absolute necessity of interface treatment which, as will be shown in the next section, can sufficiently be achieved using Mesh 3.

6.2.1   Effect of Shearing Gas Flow

To investigate the consistency of the interface treatment technique with increasing gas flows, two gas velocities reported in table 6.2 are simulated. A quantitative and qualitative analysis is performed and the reliability of the CFD simulations is discussed. After validation, the simulation results are used to conduct a flow field analysis showing the unlimited possibilities and useful insights into the multiphase phenomena which are either impossible or only possible with high physical and financial efforts associated with the corresponding experimental investigations.

Quantitative Analysis

In table 6.2, the time averaged momentum transferred to the film is compared with the experimental data. It can be seen that the predicted momentum transfer is almost 10 times the experimental value when no interface treatment is used. Keeping in mind the tolerance in the measuring instrumentation $(\pm 1Pa)$ and the sources of error in the CFD simulations, it can be said that the results produced with the interface treatment technique converge very well to the experimentally determined values.

Gas Flow Rate $[kg/s](u_{GS}[m/s])$	Additional Momentum Transfer to Film$[Pa]$		
	Experimental	Without Interface Treatment	With Interface Treatment
0.1 (8.79)	6.41	55.92	3.70
0.125136 (11)	12.43	93.52	6.89

Table 6.2: Comparison of experimental and calculated momentum transfers (Horizontal, co-current flow with pure water; $\dot{V}_L = 8.34l/min$)

Qualitative Analysis

Figures 6.5a-b show the instantaneous iso-surface of 50% liquid volume fraction without (top) and with (bottom) interface treatment corresponding to the mean shearing gas flows of ~ 9 and $11m/s$. The time step is such that the characteristic appearance of the film surface (waveforms) can be considered as quasi periodic i.e. recognizable in all later time steps. This, as mentioned earlier, usually happens after the film completes one pass of the test section and takes about $3secs$. The iso-surface is coloured with the height coordinate in meters with the minimum $(0.2m)$ positioned at the channel's bottom (the channel's floor). It can be seen that no surface disturbances can be reproduced on the film surface without interface treatment (see Figs. 6.5a-b (top)). On the other hand, a completely different surface topology results with interface treatment (Figs. 6.5a-b (bottom), see also Fig. 6.6) where the film surface remains covered with small amplitude three dimensional waves. The lack of squalls/pebbles on the film surface, unlike the experimental observations (Fig. 4.8), can be a direct result of inadequate streamwise mesh resolution ($3mm$, see table 6.1); hence, the underpredicted momentum transferred to the film can be explained. An interesting point to be noted here is that the wave development length also decreases with an increase in the mean shearing gas flow from ~ 9 to $11m/s$ (compare Figs. 6.5a-b (bottom)), a physically plausible phenomenon which is also observed in the experimental investigations (Fig. 4.8). The wave generation mechanism as concluded from the experimental observations in section 4.2.3, i.e. a single 2D wave which spreads along the width, distorts and divides into a number of smaller waves, is also fully recognisable in the two figures with interface treatment. In addition to that, another experimentally observed phenomenon like an increase in the wave amplitude and frequency with an increase in the shearing gas flow can also be confirmed by a careful comparison between Figs. 6.5a-b (see also Fig. 6.6).

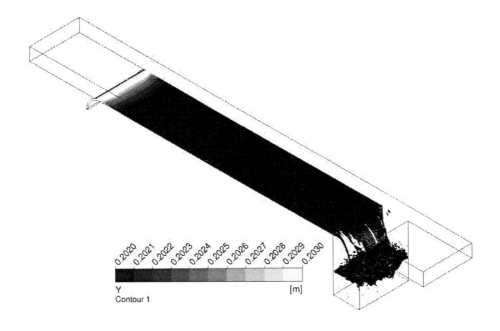

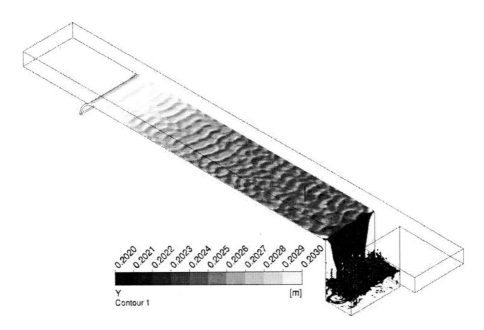

Fig. 6.5a: 50% iso-surface without (top) & with (bottom) interface treatment
$(u_{SG} = 8.79m/s, \dot{V}_L = 8.34l/min)$

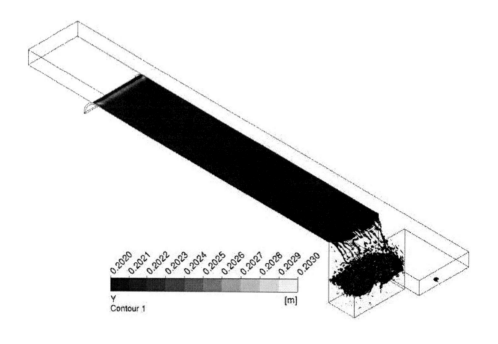

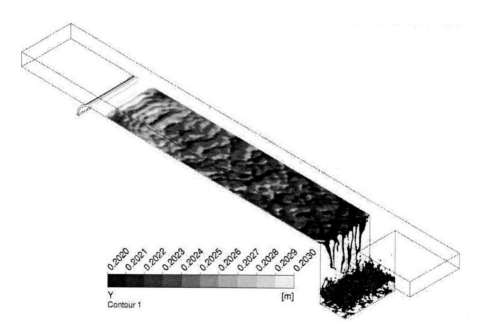

Fig. 6.5b: 50% iso-surface without (top) & with (bottom) interface treatment
$(u_{SG} = 11m/s,\ \dot{V}_L = 8.34l/min)$

Another global consequence of the interface treatment can be seen by plotting the width centred film thickness along the channel length as shown in Fig. 6.6. The film thickness calculated without interface treatment is less than $2mm$, which is in general, almost a millimetre less than the film thickness calculated with the interface treatment technique. Moreover, without interface treatment, an increase in the gas flow results only in a decrease in the mean film thickness. In the case with interface treatment, the increase in shearing gas flow is also associated with a decrease in the wave development length and an increase in the wave amplitude.

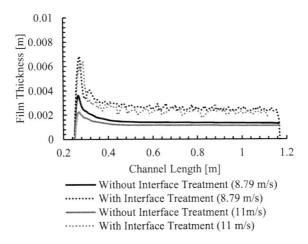

Fig. 6.6: Influence of the gas flow on the width centred film thickness distribution

Flow Field Analysis
The effect of interface treatment on the flow field can now be qualitatively analysed and discussed in relation to the expected behaviour of the turbulence modelling, especially across the gas-liquid interface, and the resulting velocity profile.

Figures 6.7a-b give the instantaneous (last time step) streamwise velocity distribution on a width centred plane. When the air meets the liquid film in the test section, the air boundary layer on the film surface grows rapidly to more than half the channel's height due to the unphysical interaction between the phases (false momentum transfer) if the interface treatment is not used (see figure 6.7a-b (top)). This results in an unphysical velocity peak towards the upper wall of the channel (see also Fig. 6.8) for the same reason as explained in section 5.3.2. The magnitude of the velocity peak above the film is also considerably large as compared to the almost uniformly distributed velocity profile with interface treatment. To further elaborate the turbulent flow field, a local analysis of the instantaneous (last time step) velocity distribution and the corresponding turbulent quantities in both phases is performed at a width centred location $1m$ from the air inlet which is shown by a solid vertical line in Figs. 6.7a-b. It can be seen from Fig. 6.8 that in both cases $(0.1$ and $0.125136kg/s)$, as expected, there is a huge production of turbulent kinetic energy resulting in a very high turbulent viscosity in the vicinity of the gas-liquid interface. Consequently, there is an unphysical gas side velocity profile and immense pressure drops (see Table 6.2) pointing out the absolute necessity of the interface treatment techniques for the VOF method to be applicable to shear driven films. It should be noted at this point that due to the transient nature of the flow field, the turbulent quantities at the gas-liquid interface continuously change in time and space; therefore, the information provided in Fig. 6.8 is instantaneous and can only be used to appreciate the absolute necessity of the interface treatment e.g. to compare with the qualitative wall behaviour provided in Figs. 5.4-5.6.

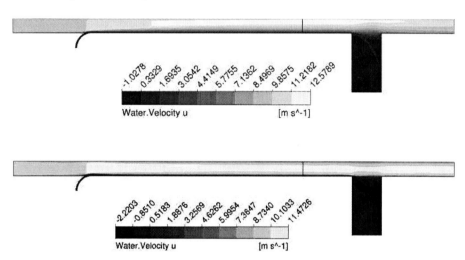

Fig. 6.7a: Streamwise velocity distribution on a width centred plane ($u_{SG} = 8.79 m/s$) (Without (top) & with (bottom) interface treatment)

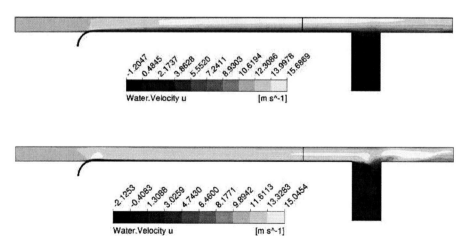

Fig. 6.7b: Streamwise velocity distribution on a width centred plane ($u_{SG} = 11 m/s$) (Without (top) & with (bottom) interface treatment)

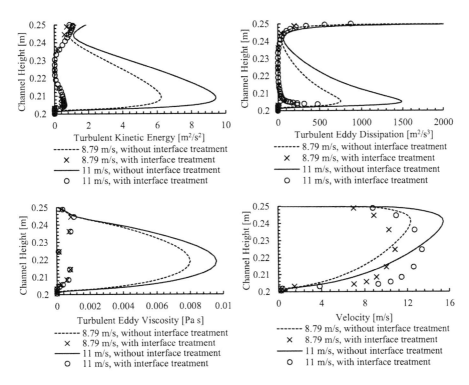

Fig. 6.8: Local flow field analysis at a width centred line $1m$ from the air inlet

6.2.2 Effect of Gravity

The response of the interface treatment technique to the gravitational force is investigated by keeping the air velocity and liquid flow rate constant at $8.79m/s$ and $8.34l/min$ and modifying the gravitational vector to simulate the experimental inclinations of 5 and $10°$. A point to be noted here is that the VOF method utilizes a homogenous turbulent field for both phases i.e. if simulated turbulent, both phases will be considered as turbulent. However, in the case with considerable gravitational influence e.g. the case with $10°$ inclination, the shearing gas flow and the gravitational induced additional momentum drives the film quickly into the laminar regime (see section 4.2.3). Therefore, unlike the gas flow, the film flow must be considered as laminar and a provision is necessary to implement this behaviour in the homogenous field VOF model. In this work, it is achieved with the help of the same source term meant for the interface treatment but this time extending it to the complete film domain. This is physically appropriate because the source term meant for the interface treatment is doing nothing else than to dissipate the entire turbulent kinetic energy calculated by the turbulence model to give a zero turbulent viscosity. Hence, the effective viscosity calculated by the solver will comprise only of the laminar viscosity and therefore, the film can be simulated as laminar independent of the turbulent gas flow. Nevertheless, a small discrepancy will remain due to the turbulent wall function at the wall where the film is flowing.

Quantitative Analysis

Table 6.3 reports the CFD predictions of the momentum transfer with and without interface treatment and the corresponding values recorded during the experimental investigations. As mentioned in the paragraph dealing with the sources of error (see pg. 104), for higher values of numerical diffusion, the pressure drop with a film in the test section is expected to be significantly underpredicted. With the gravitational induced additional momentum, the film from the test section falls in the form of a liquid sheet in the separator tank, hence, worsening the level of numerical diffusion. This can directly be seen from Figs. 6.9a-b, where the interface could not be resolved soon after the test section. For the $5°$ case, the experimental momentum transfer can still be reached fairly well. For the $10°$ case, however, the numerical diffusion causes in average more than $70g/s(\sim 45\%)$ of liquid to exit from the air outlet preventing a direct comparison with the experimental momentum transfer. The unphysical fluid is denser than air alone and is therefore, further accelerated in the outlet module due to the action of gravitational forces. In this manner, the pressure drop with a film 'ΔP_{WF}' in the test section becomes smaller than the pressure drop without a film 'ΔP_{WoF}' in the test section, hence, a negative value results.

| | Additional Momentum Transfer to Film$[Pa]$ | | |
Inclination Angle	Experimental	Without Interface Treatment	With Interface Treatment
$5°$	3.54	49.56	1.51
$10°$	1.2	45.14	-7.95

Table 6.3: Effect of gravity on the momentum transfer
(Co-current flow with pure water; $u_{GS} = 8.79 m/s$, $\dot{V}_L = 8.34 l/min$)

Qualitative Analysis

For inspecting the behaviour of the film surface, 50% iso-surface of the liquid volume fraction is provided in Figs. 6.9a-b. It can be seen that without interface treatment no surface instabilities can be observed on the film surface which contradicts the experimental observations (see Fig. 4.10). On the contrary, surface instabilities can be recognized on the film surface with interface treatment. However, a direct comparison with the experimentally observed wavy structure cannot be achieved. This can be explained based on the fact that due to the action of gravitational force, the film thins and the wavy structure becomes finer. To resolve the even finer waveforms than in the horizontal case, finer interface resolution is required; yet, the cell resolution used here is just enough to achieve a global validation. Therefore, a direct resemblance to the experimental observations cannot be expected. A closer look at the width centred film thickness distribution shown in Fig. 6.10, however, confirms the general behaviour of film under the influence of gravity when compared with the *Reference Case*[5] (without gravity). It can be qualitatively confirmed from experiments that with increasing inclination angles, the film becomes thinner and the development length for the wave generation becomes larger. In the case with interface treatment, the mentioned trend is clearly visible whereas it is absent in the case without interface treatment.

---

[5]Reference Case: horizontal, co-current flow with pure water; $u_{SG} = 8.79 m/s$, $\dot{V}_L = 8.34 l/min$

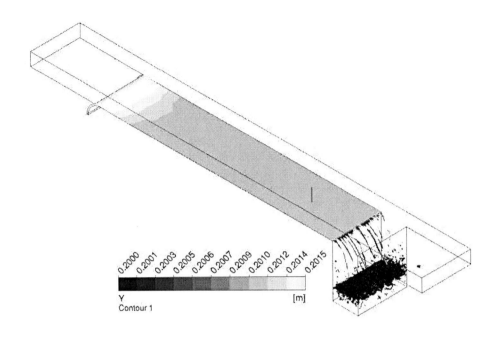

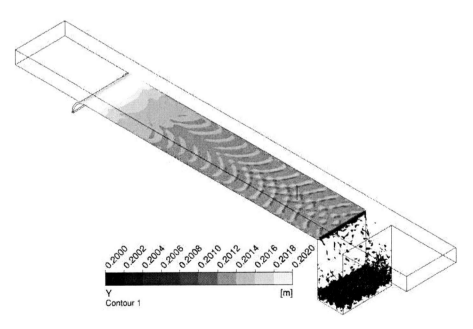

Fig. 6.9a: 50% iso-surface without (top) & with (bottom) interface treatment
$(5°,\ u_{SG} = 8.79m/s,\ \dot{V}_L = 8.34l/min)$

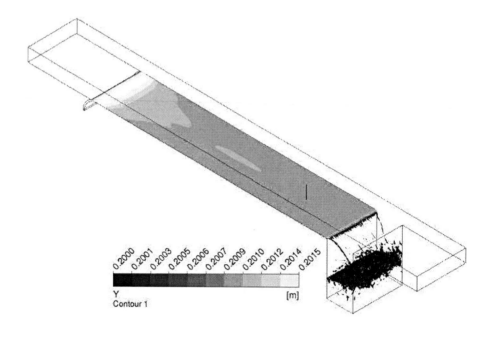

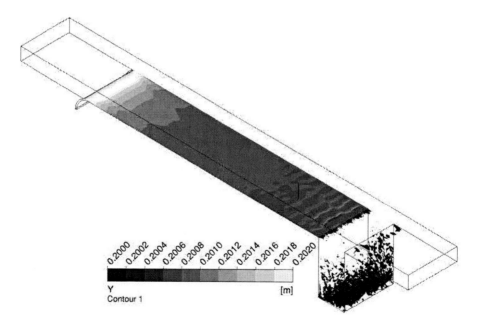

Fig. 6.9b: 50% iso-surface without (top) & with (bottom) interface treatment
(10°, $u_{SG} = 8.79 m/s$, $\dot{V}_L = 8.34 l/min$)

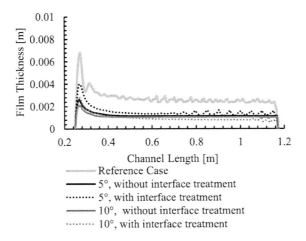

Fig. 6.10: Influence of gravity on the width centred film thickness distribution

6.2.3   Effect of Film Properties

In many engineering situations, the film properties may change during operation. The topic of interest in this thesis "oil film dynamics in bearing chambers" also encounters a considerable change in the oil properties with the operating temperatures. The change in oil properties like surface tension and viscosity strongly influence the film dynamics. Therefore, it is important to investigate the response of the interface treatment technique to the change in physical properties of the film. Similar to the experimental investigations, first, the influence of individual film properties on the film dynamics is simulated and compared with the experimental results. In the next step, the simultaneous change in the film physical properties is simulated and compared with the experimental results. After a successful global validation, the waveforms appearing on the film surface and the general trend of the film thickness along the channel length is analysed in accordance with the corresponding experimental observations.

Quantitative Analysis

The global quantitative analysis is presented in table 6.4. In all the simulated cases, a fairly well agreement with the experimental results is evident.

Film Properties	Additional Momentum Transfer to Film $[Pa]$	
	Experimental	With Interface Treatment
Low surface tension $(36.5mN/m)$	3.01	4.80
High viscosity $(5cSt)$	7.05	5.93
Low surface tension and High viscosity	5.70	6.67
Reference Case	6.41	3.70

Table 6.4: Effect of film properties on the momentum transfer

Qualitative Analysis

The qualitative analysis revealed that with the decrease in surface tension, the simulation predicts waveforms similar to the reference case as can be seen in Figs. 6.11-6.12. This contradicts the experimental observation (see Fig. 4.12a) where a relatively flat surface results. However, based on the literature review conducted in this thesis (see chapter 2 for details), the role of surface tension on the generation of surface waves is contradictory and cannot be established yet. The clarification of this question requires further experimental and numerical investigations.

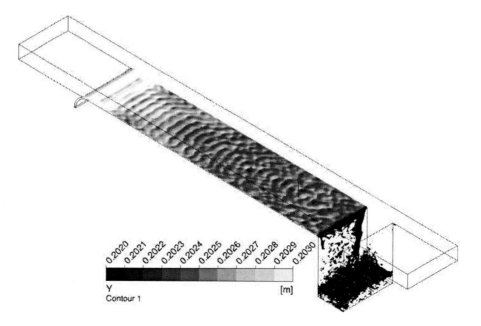

Fig. 6.11: 50% iso-surface - Effect of low surface tension on the film surface ($\sigma = 36.5mN/m$, $u_{SG} = 8.79m/s$, $\dot{V}_L = 8.34l/min$)

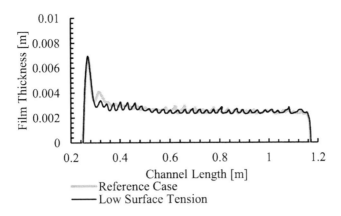

Fig. 6.12: Influence of reduced surface tension on the width centred film thickness distribution

From the experimental observations, in general, it can be said that the increase in viscosity results in film thickening and a delay in wave generation i.e. an increase in the development length (see Fig. 4.14a). The waves that appear are, however, longer in wavelengths and higher in amplitudes. This behaviour can also be observed from Fig. 6.13-6.14 where the CFD predictions are presented.

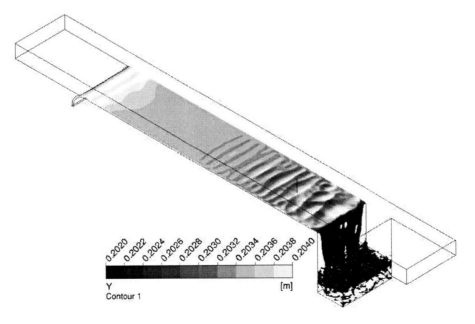

Fig. 6.13: 50% iso-surface - Effect of high viscosity on the film surface
$(\mu = 5cSt,\ u_{SG} = 8.79m/s,\ \dot{V}_L = 8.34l/min)$

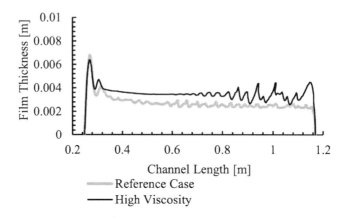

Fig. 6.14: Influence of high viscosity on the width centred film thickness distribution

The influence of the simultaneous decrease in surface tension and increase in viscosity on the film surface structure results in a superposition of both effects. The experiments revealed that viscosity plays a dominant role in the wave generation mechanism (see Fig. 4.16a), consequently, the waveforms resemble the case with high viscosity only which is also predicted by the CFD simulation as can be seen in Figs. 6.15-6.16.

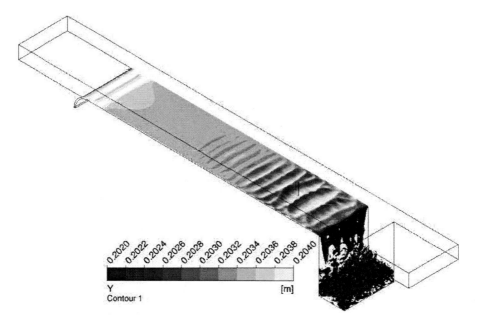

Fig. 6.15: 50% iso-surface - Effect of low surface tension & high viscosity on the film surface
$(\sigma = 36.5mN/m, \mu = 5cSt, u_{SG} = 8.79m/s, \dot{V}_L = 8.34l/min)$

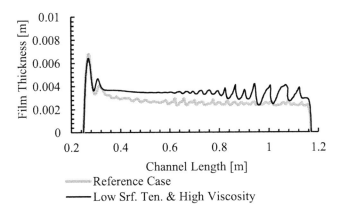

Fig. 6.16: Combined influence of the reduced surface tension & high viscosity on the width centred film thickness distribution

All test cases presented here produce satisfactory results for the engineering applications. The influence of viscosity can be reproduced fairly well whereas the influence of surface tension shows a false tendency; yet, the results remain in the right orders of magnitude in both cases. Moreover, the superposition of properties also yields the right tendency and indicates a dominant role of viscosity which is in agreement with the experiments. The numerical investigations presented in this chapter can now be concluded with the remark that interface treatment is proved to be absolutely vital for the VOF model to properly deal with the shear driven film flows. The VOF model in its original form is not applicable to such flows and if applied, will lead to serious discrepancies.

7 Application Oriented Test Case - ITS HSBC Test Rig

In this chapter, the interface treatment technique is employed to successfully validate a model of an engine representative geometry including all the salient features like a rotating shaft, a labyrinth seal and a scavenge pump. Only the simplified interface treatment can be applied due to the lack of proper implementation of the refined interface treatment at the time of simulations. The simulation methodology, given by Peduto et al. (2011), assumes that the scavenge system can be sufficiently analysed by simulating only the lower half ($90-270°$, see Fig. 2.1) of bearing chambers. The necessary experimental data for validation comes from the experiments conducted by Peduto et al. (2010) using the ITS HSBC test rig. Before going into the details of the simulation setup, in the following, the experimental setup and the non-dimensional validation quantities are briefly introduced.

7.1 Experimental Setup

In Fig. 7.1 (Left), an axial cross-section of the two bearing chambers in the ITS HSBC test rig is shown. The experimental investigations were carried out in the bearing chamber number II (see also Fig. 2.1) with a height to width ratio of 0.066. The diameter of the vent and scavenge off-takes was $10mm$. An engine oil of type "Jet Mobile II" was pumped at the rate of $150l/hr$ and a small sealing air flow rate of $3g/s$ was kept through the labyrinth seals to avoid oil leakages. To avoid droplets in the system, the $\varnothing128mm$ rotor was rotated at a small rotational speed of $3500rpm$. Experiments were performed at ambient conditions.

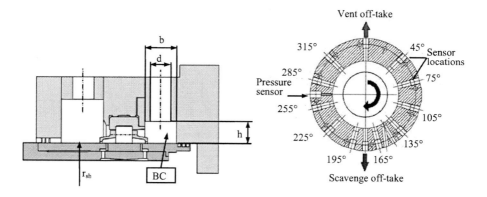

Fig. 7.1: Investigated bearing chamber (Left); Chamber probe locations (Right)

To determine the circumferential film thickness distribution, the chamber outer wall was equipped with 10 capacitive sensors at $30°$ intervals as shown in Fig. 7.1 (Right). Due to a small width of the chamber, the axial film thickness could not be resolved. The employed capacitive sensors have a diameter of $\varnothing10mm$ and give the mean film thickness over the sensor area. A detailed explanation of the measurement technique applied in the investigations can be found in Gorse et al. (2004).

The comparison between the results of the CFD simulation and the experimental data are performed using non-dimensional quantities which are defined as follows:

The scavenge ratio 'SR' describes the efficiency of a scavenge system in aero engine bearing chambers and is defined as

$$SR = \frac{\dot{V}_{total,pump}}{\dot{V}_{oil,in}} \tag{7.1}$$

The predicted and measured film thickness is non-dimensionalized with the help of the maximum film thickness achieved in either case.

$$h^*_{f,angle} = \frac{\overline{h}_{f,angle,EXP/CFD}}{\overline{h}_{max,angle,EXP/CFD}} \tag{7.2}$$

For a global validation, a non-dimensional parameter can be achieved by taking the ratio of predicted to the measured oil discharge from the scavenge off-take (surface '3' in Fig. 7.2).

$$\dot{m}^*_{oil} = \frac{\overline{\dot{m}}_{oil,pipe,CFD}}{\overline{\dot{m}}_{oil,pipe,EXP}} \rightarrow 1 \; for \; best \; comparison \tag{7.3}$$

7.2 Simulation Setup

CFD Domain

As shown in Fig. 7.2, the CFD domain consists of the lower half of the bearing chamber.

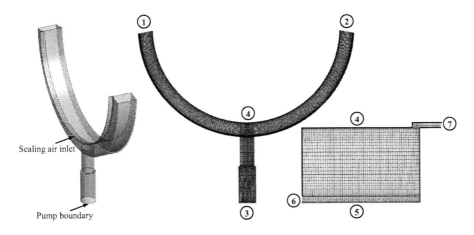

Fig. 7.2: CFD domain and boundary conditions

In chapter two, it is shown that in the upper half of the bearing chambers (near the vent port) the circumferential distribution of the cooling oil film is usually a millimetre or less, whereas near the scavenge system the film thickness can be as high as $5mm$. In contrast to thick films, thin films are usually very stable and analytically describable (see chapter 2 and 4 for details). Hence, the upper half of the bearing chambers can be predicted with other less expensive methods (e.g. thin film models). As applying a full 3D numerical model to resolve the complete bearing chambers will result in large computational and time efforts due to prohibitly large number of cells required. Therefore, to keep the number of cells in a limited range and capture only the area of interest, a new methodology is adopted by Peduto et al. (2011). With the help of an open channel boundary condition, it is shown that the scavenge system can be sufficiently analysed by simulating only the lower half of the bearing chambers. Nevertheless, the methodology can also be used for a global analysis of the upper half of the bearing chamber in a sense that the amount of air/oil leaving the chamber through the vent line can be calculated indirectly from the known amount leaving through the scavenge line.

Boundary Conditions

Since only half of the bearing chamber is simulated, the boundary surfaces '1' and '2' in Fig. 7.2 act as simultaneous inlets and outlets. This means that depending on the flow field which establishes with the help of other boundary conditions (surfaces '3', '4', '6' & '7'), a suitable mass flow of both phases independent of each other can either enter or exit the surfaces '1' and '2' simultaneously. This is achieved with the help of a total pressure boundary condition combined with the open channel option. This boundary condition requires the knowledge of the film thickness which is provided from the experimental measurements ($h_{f,105°}^{*} = 0.46$ & $h_{f,255°}^{*} = 0.64$). However, during the simulations, it is noticed that the knowledge of the film thickness does not need to be very accurate. Hence, it can be said that a reasonable guess of the film thickness should be enough for the presented methodology.

The scavenge pump is simulated by assigning a negative velocity inlet to the end of the scavenge line (surface '3' in Fig. 7.2). From several preliminary numerical experiments, it could be established that such a boundary condition behaves in the desired manner and does not differentiate between the phases. The test case is investigated with a scavenge ratio 'SR' of '4'.

The rotating shaft (surface '4' in Fig. 7.2) is defined as a moving wall with a rotational speed of $3500rpm$. Oil enters the domain through the side wall (surface '6' in Fig. 7.2) of the chamber with an assumed height and an inlet swirl of $45°$ in the direction of shaft rotation, modelling the oil flow from a roller bearing. Gorse et al. (2008) showed in his experiments that almost no droplets were generated at the roller bearing unless a strong cross flow of air was artificially introduced through the bearings. Moreover, several numerical experiments have revealed that the inlet film thickness has no influence on the results as long as the film thickness is smaller than the film thickness in the chamber and the inlet swirl ranges between 0 & $45°$. Therefore, introducing the oil in the chamber as a film can be considered as a reliable and easy to implement boundary condition which is also very close to the actual experimental scenario.

Sealing air is introduced in the domain through the labyrinth duct (surface '7' in Fig. 7.2). The velocity profile at the sealing air inlet is calculated separately in a 2D axisymmetric labyrinth seal model to ensure an appropriate profile. Schramm et al. (2004) showed that a 2D axisymmetric approach is accurate enough to resolve the flow field in similar labyrinth seals. Accordingly, an inlet swirl of about $45°$ is calculated with the labyrinth seal model. It has to be mentioned here that the inlet swirl of the sealing air has a large influence on the air flow in a bearing chamber. For example, depending on the inlet swirl of the sealing air, the toroidal vortices in the chamber can change in number and direction of rotation. Therefore, correct sealing air flow inlet conditions are absolutely necessary for a plausible two-phase flow analysis inside a bearing chamber.

Mesh Independency

Three different mesh densities, shown in table 7.1, are used to test the fidelity of the introduced methodology. The simplified interface treatment given by Eq. 5.36 in chapter five is implemented in all cases. The mesh is refined at the expected location of the gas-liquid interface in three stages. The time and area-averaged film thickness at a circumferential position of $165°$ and the global oil flow rate discharged through the off-take are analysed. It can be seen from table 7.1 that refinements from Mesh 1 to Mesh 3 have led to only small discrepancies ($< 8\%$). A point to be mentioned here is that due to the complexity of the simulations, the CFD code adopts a very small time step ($\sim 1 \times 10^{-5} - 1 \times 10^{-6} sec$) to keep a constant Courant number of 1. On the other hand, this chapter only aims to show that the derived interface treatment technique is also applicable to application oriented test cases. Therefore, to reduce the computational time and efforts, Mesh 1 is selected to carry out the numerical simulations of the investigated test case.

Mesh	No. of Elements	Compared Quantities	
		$h^*_{f,165°}$	$\dot{m}^*_{oil}$
Mesh1	506,716	0.69	0.91
Mesh2	749,028	0.74	0.94
Mesh3	1,065,342	0.72	0.93

Table 7.1: Mesh Independency

7.3 Validation and Analysis

As mentioned earlier, a transient simulation is performed with an adaptive time step control. The simulation is started first without rotating the shaft and without the sealing air entering the chamber until the liquid (oil) reaches the scavenge pipe (free draining). This took about $0.2secs$. The shaft is then rotated and the sealing air is introduced in the chamber. The simulation is continued until the flow field reaches a quasi-steady state. The test case is then validated against local and global quantities gained from the experiments. After the validation, the important aspects of the flow field are discussed with the help of the CFD results.

In Fig 7.3, the non-dimensional film thicknesses at six angular positions are compared. The predicted (CFD) film thicknesses, similar to the experimental film thickness, are also averaged and non-dimensionalized (see Eq. 7.2). It can be seen that the film thickness distribution without interface treatment shows a large deviation from the experimental distribution. Especially, from the scavenge off-take, in the direction of shaft rotation between $225-255°$, the film thickness decreases to nearly zero leading to a dry patch in the simulation which was not observed in the experimental investigation. From $105°$ onwards, in the direction of the scavenge off-take and beyond, the film distribution is overestimated in the case without interface treatment. This behaviour of the film distribution can be explained by keeping in mind the unphysically high pressure losses that the air flow encounters due to the unphysically high momentum transfer to the oil film in the case without interface treatment. Accordingly, the air flow takes the route of less resistance which for the investigated boundary conditions is mostly out of the domain. Therefore, in the case without interface treatment, only the counter-current regime establishes on either side of the scavenge off-take; hence, the rotating shaft enhances the air flow in one half $(180-255°)$ and reduces it in the other $(105-180°)$. In the half where the air flow is enhanced, complete flow reversal results, hence, the occurrence of a dry patch can be explained. Similarly, due to the deceleration in the half where the air flow is reduced the occurrence of a local film thickening can be explained.

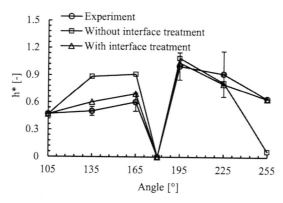

Fig. 7.3: Film thickness distribution without and with interface treatment

On the other hand, the film thickness distribution in the case with interface treatment remains within $\pm15\%$ of the measured film thickness distribution. The film thickness is slightly underpredicted between $180-255°$ and slightly overpredicted between $105-180°$. This, as explained in section 5.4.2.1, can be a direct result of a slight imperfection in the simplified interface treatment method employed in this study. The improved agreement between the predicted and the measured data can be attributed to the minimized additional (false) momentum transfer to the oil film. As a result, the air velocity gradient at the interface is automatically corrected and the momentum is transferred by the interfacial shear stresses and not by turbulent stresses.

Regarding the global validation, the VOF model without interface treatment results in 80% of the oil discharge whereas with interface treatment in 91% from the scavenge off-take when compared to the experimental value. Without interface treatment, the lower discharge from the scavenge pipe can again be explained on the fact that on either side of the off-take counter-current flow regime exists. Hence, the air flow restricts the oil film entering the chamber at the surfaces '1' and '2' $(105°/255°)$. In the near off-take region, the air flow holds the wall film in the process of thickening until the gravitational force dominates and pulls the film towards the off-take. The interface treatment technique minimizes the unphysical momentum transfer, hence, the oil discharge mass flow through the scavenge off-take is significantly improved.

Qualitatively, it can be said that certain flow features like the recirculation region can only be captured with interface treatment as shown in Fig. 7.4 (Right) which is completely absent in the case without interface treatment as shown in Fig. 7.4 (Left). Besides capturing the recirculation region, it is also found that the occurrence of recirculation is a highly dynamic phenomenon which will be dealt in detail later in this chapter. Generally, the film surface remains flat for the case without interface treatment whereas other small instabilities (waves) are also evident with interface treatment. In contrary to the case without interface treatment where a complete dry out is predicted, no large dry out (only small points) is predicted in the case with interface treatment which is also in agreement with the experiments.

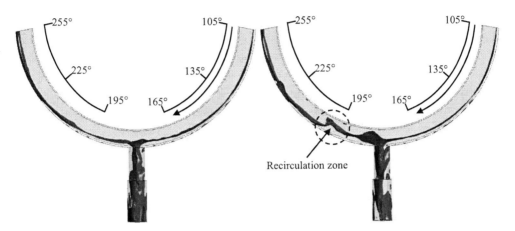

Fig. 7.4: 50% iso-surface of oil volume fraction without (Left) & with interface treatment (Right)

Flow field Analysis
As mentioned earlier, the recirculation zone cannot only be resolved with the help of improved turbulence modelling (interface treatment) but it is also found that the flow in this area is highly dynamic. This means that the recirculation zone cannot be localized to a certain position in time or space. Instead, it seems that as a result of interaction between the driving forces and off-take flows (geometry and scavenge ratio) it evolves at the off-take, carried away with the shearing air flow and in the process accumulates mass to a point where gravity balances the shear forces. Simultaneously, during the process of accumulating mass, the recirculation zone reduces the flow area for the sealing air; hence, builds up a huge resistance to the air flow and results in a split of the air flow. A major part of air now flows in the opposite direction (see Fig. 7.5 (Right) at $t = 0.41s$) whereas the rest still flows over the top of the recirculation zone and is accelerated. The top of the recirculating oil is teared off into large ligaments as a result of local acceleration until the recirculation region completely disappears. This leads to the change of air flow in the original direction and the flow now resembles the state at a previous time step (see Fig. 7.5 (Left) at $t = 0.28s$). From there onwards, the flow phenomenon repeats itself and it seems that its nature is periodic for the investigated conditions. Unfortunately, a direct comparison to the test data is not yet possible. However, from the counter-current investigations presented in chapter 4, the conclusion can be drawn that the dynamic behaviour of the recirculation zone explained here is plausible. In chapter 4, it is shown that with increasing counteracting air flows big solitary waves

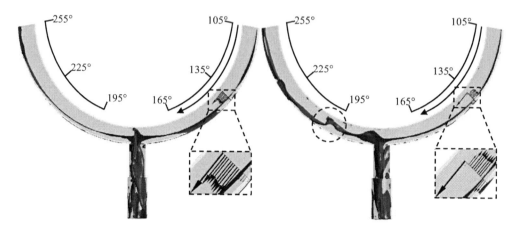

Fig. 7.5: 50% iso-surface of oil volume fraction at $t = 0.28s$ (Left) and $t = 0.41s$ (Right) showing a transition from co- to counter-current flow regime with time

of periodic nature form on the surface of liquid film and roll in the direction of air flow whereas a part of film still flows in the original direction i.e. opposite to the air flow. With further increase in the air flow, the solitary waves travel on the film surface with ligaments tearing off from the wave peaks as predicted by the CFD simulation presented in this chapter.

To investigate this phenomenon in the experimental data, the raw film thickness signals are processed to look for the evidence of periodicity. However, it should be noted here that periodicity can only be captured if the frequency of periodicity is less than half the frequency of the measuring equipment. Accordingly, the film thickness signals at all measured locations are investigated. Indeed, periodically fluctuating signals are found for all locations. To further elaborate this fact, the raw signal of the non-dimensionalized film thickness at $195°$ is provided in Fig. 7.6 (Left).

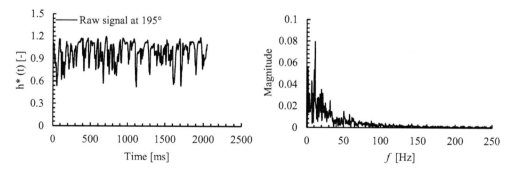

Fig. 7.6: Raw signal of the measured film thickness at $195°$ (Left); Corresponding Fast Fourier Transform (Right)

The time trace of the signal already reveals a periodic nature of the signal. To further analyse the frequency content, a FFT analysis of the signal is performed on the raw film thickness signal (Fig. 7.6 (Right)). From the result of the FFT, a dominant frequency of $11Hz$ can clearly be recognized.

Therefore, it can again be said that the prediction made by CFD is physically plausible. However, as mentioned before, it is too early to state anything especially as the CFD simulation is run only for about $0.5s$ whereas the FFT shown is performed on a data of two seconds.

Another important and interesting phenomenon predicted by the CFD simulation is the existence of strong secondary flows which can also be traced back to the experimental investigations of Gorse et al. (2003) in a similar bearing chamber geometry. CFD reports secondary flows with a velocity up to 70% of the core velocity. This indicates a variation in the axial film thickness distribution which can also be seen in the CFD prediction (Fig 7.7).

Fig. 7.7: Air streamlines coloured with velocity on an axial section at $195°$

Based on the excellent quantitative and qualitative comparison of the CFD predictions with the experimental data, it can now be stated that the VOF model with interface treatment offers a huge potential for simulating oil film flows in bearing chambers. With the help of suitable sub-models, the methodology presented here can be extended to simulate droplet stripping and droplet-film interaction phenomena occurring in bearing chambers, hence, paving the way for more engine relevant simulations.

8 Summary and Outlook

The oil flow in bearing chambers plays a vital role of fulfilling the tasks such as cooling and lubricating, and cleaning the foreign particles which might be deposited in bearing chambers. Most commonly, the oil is introduced in the form of a jet and predominantly flows in the form of a film on the chamber housing and thereby cooling the chamber walls as well. To avoid any oil leakage to the hot surroundings, the bearing chambers are usually sealed with the help of contactless labyrinth seals pressurized by sealing air taken from the compressor bleeds. The fast moving shaft accelerates the air in the circumferential direction and in the process influences the oil film dynamics inside the bearing chambers. This interaction between the oil film and the sealing air can adversely affect the cooling efficiency by increasing the residence time of oil inside a bearing chamber. This is an extremely important issue for all recent and future trends in aero engines which require an increase in the temperature limits. This in turn means a strict reliance on the development of better components for the heat management system. Therefore, for innovative designs and optimization of bearing chambers with respect to the new challenges, the oil film dynamics need to be known in advance. Considering this fact, a number of researchers have carried out experimental investigations in the past using test rigs with some degree of abstraction of the bearing chambers. Regarding the film dynamics, circumferential film thickness distributions and sometimes, the film velocity profile at a point were reported. The analysis revealed that for the most part, the bearing chamber typical film is relatively thin ($\leq 1mm$) and stable and can be fairly-well predicted with the available thin film models. In the vicinity of the scavenge off-takes, however, the film can grow to a couple of millimetres ($\sim 2-5mm$) and is highly dynamic in nature. Further insights into the dynamics of thick films, due to the limitations on the multiphase measuring techniques, optical access and coexisting flow phenomena (e.g. droplet-wall film interactions, off-take disturbances, curvature & oil feed from all possible circumferential positions), are very difficult to achieve in complex bearing chamber test rigs. This can, however, be done with relatively low financial and human efforts if the thick film flow phenomena can be resolved using modern CFD techniques. Therefore, in the framework of this thesis, the fundamental physical understanding and CFD modelling of the near scavenge thick film flow phenomena are set as the primary objectives.

The film boundary conditions are either unknown or can only be known very vaguely in complex bearing chamber test rigs which hinders any attempt to reliably correlate the film flow data. There is a growing trend in employing modern CFD techniques which also require well-defined boundary conditions. The fact that multiphase models are flow regime dependent and CFD solutions are usually very sensitive to the choice of boundary conditions, multiphase film models cannot be analysed appropriately. Moreover, the CFD technique also requires appropriate validation data which for the undeveloped nature of film flow, as mentioned earlier in section 4.1.2, is difficult to achieve from engine realistic test rigs with complex geometries. In literature several experimental studies dealing with shear driven film flows can be found but either inappropriate boundary conditions are used or insufficient knowledge on boundary conditions is provided. Apart from this, almost all data in literature is reported for fully developed film conditions (very long test rigs); hence, do not resemble near scavenge thick film flows. In section 2.1.1, with the help of a force balance on an infinitesimal film element, it is shown that the film flow in the vicinity of the scavenge off-takes can be reproduced in a linear rotatable test rig. Therefore, a simple test rig is designed and built in the scope of this thesis to provide the fundamental understandings as well as appropriate test cases of isolated film flows for CFD applications by providing all the necessary information for a reliable simulation. Although the test rig is simple, nevertheless, it represents a class of flow (oil film flow) encountered in aero engine bearing chambers. This is done by using similarity to the ITS HSBC test rig. The HSBC test rig revealed that the thick oil film may behave very differently in the co- and counter-current regime even for high shaft speeds. Accordingly, with the help of the test rig designed in the context of this thesis, both flow regimes can be investigated. To ensure engine relevant conditions the film properties, namely, surface tension and viscosity are matched to that of oil during engine operation. For a quantitative analysis, a parameter determined from the single phase pressure drop measurements (only in gas) is introduced to effectively analyse the undeveloped film flows. The complete (integral) momentum transfer to the film from the shearing air flow can be measured using the introduced parameter. In this manner, as a first step towards a more detailed local validation, an extensive global validation of the CFD model can be achieved. The fundamental know how is gathered by investigating the influence of major factors, first individually and then by superimposing, affecting the film dynamics in a typical bearing chamber environment. It is found that air experiences the oil film in both regimes as an additional resistance (in the test section); consequently, the losses are proportional to the square of superficial mean air velocity. It is also shown that the momentum losses occurring in the counter-current regime are always larger than in the co-current regime for all gas flows (all engine shaft speeds). For high gas flows (high shaft speeds in engine) the losses on the counter-current side (churning losses) can be several magnitudes larger than the co-current side (no churning losses). It is also shown that for the conditions investigated and relevant to the bearing chambers the possibility of wave formation leading to droplet shedding on the co-current side is very unlikely. Significant droplet shedding occurs in the counter-current regime. A qualitative analysis of the film surface is also performed where characteristic waveforms for the investigated regimes (co- & counter-current) are found and discussed in relation to the quantitative analysis. The quantitative and qualitative analysis of water is presented first which serves as the reference case. The influence of physical properties is discussed in relation to the reference case. As a consequence of reduced surface tension, it is observed that the waves are largely damped and the film surface can be considered as almost flat. The reason why reducing surface tension results in an almost flat surface requires detailed investigations of the local flow field which is not dealt within the scope of this thesis. In literature, no clear answer but contradictory statements are found about the role of surface tension on the wave generation mechanism with some arguing an enhanced wavy interface while others a damped interface. To address this problem, a new investigation focusing on the role of surface tension on the wave generation mechanism is suggested. Finally, based on the detailed quantitative and qualitative analysis, the possible characteristics of the oil film near the scavenge port (in the absence of coexisting phenomena) of a typical bearing chamber are outlined.

In the past, due to limited computational resources, the potential offered by numerical techniques was only exploited to a limited extent. Recent advancements in computer hardware have drastically changed the upper limits on the available computational power. Therefore, a 3D (CFD) simulation of a bearing chamber seems to be a reality today. In this thesis, first, the state of the art multiphase models are analysed for their feasibility and applicability to model shear driven thick film flows in bearing chambers. After a thorough physical understanding of the film flow phenomena in the experimental investigations of this thesis, Volume-Of-Fluid Model is shown to be the most suitable multiphase model. Preliminary simulations have, however, shown serious deficiencies in the model. The deficiencies are shown to be a result of false momentum transfer between the phases due to the homogenous turbulence modelling approach and a lack of interface physics in the model. Two solutions, namely, a simplified and a refined interface treatment are suggested to improve the Volume-Of-Fluid model. The refined interface treatment is a mathematically closed technique, derived from the transport equations of Two-Equation turbulence models '$k-\varepsilon$' & '$k-\omega$', and does not require any input. The enhanced VOF model is successfully validated, first, against the test rig presented in this thesis simulating seven test conditions. They included the analysis of the effect of shearing gas flow, effect of gravity and the influence of film physical properties. An overall good agreement with the experimental momentum transfer is achieved in almost all cases. However, unlike the experimental observations, the case with reduced surface tension demonstrates a wavy interface similar to the reference case with slightly lower amplitude waveforms. Since a fairly well global validation is achieved in this case also, it can be assumed that the waves are present but probably finer that the camera eye can capture and irrelevant in the overall momentum transfer. Second, the enhanced VOF model is applied to the ITS HSBC test rig and compared with the original VOF model. According to Peduto et al. (2011), only the lower half of the bearing chamber is simulated. Peduto et al. (2011) showed that by assigning an open channel type boundary condition to the junction surfaces between the lower and upper halves of a bearing chamber, the junction surfaces act as simultaneous inlets and outlets for the phases. This has the advantage that a very thin film which is present in the upper half does not need to be resolved and hence, keeps the number of elements in a reasonable range which would have been prohibitly large otherwise. The simulation results showed that the interface treatment is absolutely necessary for reproducing typical film surface instabilities e.g. waves, recirculation zone etc... which were completely absent in the case without interface treatment. The CFD results also showed that the recirculation zone is highly dynamic and probably periodic in nature which may influence the air flow by completely changing its direction. The periodicity of the flow phenomena can also be confirmed in the FFT analysis of the time dependent film thickness signal acquired from the experiments. Therefore, the presented interface treatment technique offers the potential for simulating the film flow phenomenon in bearing chambers. At high pressures and shaft speeds, droplets are also present in the system. Since no regime specific assumption is made in deriving the interface treatment techniques it can be said that the presented interface treatment techniques can also handle droplets in the system. For the simulation of the lower half of the bearing chamber presented here, this means that if the junction surfaces can be coupled with some suitable sub-models exchanging information with the upper half, the complete bearing chamber can be simulated. Hence, it can now be stated that in terms of the numerical modelling of bearing chambers, the work conducted in this thesis lays the first milestone by providing a working tool which can model one of the major flow phenomena existing in bearing chambers.

References

Abe, Y., Akimoto, H. and Murao, Y. (1991): *Estimation of Shear Stress in Counter-Current Annular Flow*. Journal of Nuclear Science and Technology, Vol. 28(3), pp. 208–217.

Agrawal, S.S, Gregory, G.A and Govier, G.W. (1973): *An Analysis of Horizontal Stratified Two Phase Flow in Pipes*. Canadian Journal of Chemical Engineering, Vol. 51, pp. 280–286.

Akai, M., Inoue, A., Aoki, S. and Endo, K. (1980): *A Co-current Stratified Air-Mercury Flow with Wavy Interface*. International Journal of Multiphase Flow, Vol. 6, pp. 173–190.

Akai, M., Inoue, A. and Aoki, S. (1981): *The Prediction of Stratified Two-Phase Flow with a Two-Equation Model of Turbulence*. International Journal of Multiphase Flow, Vol. 7, pp. 21–39.

Andreussi, P. and Persen, L.N. (1987): *Stratified Gas-Liquid Flow in Downwardly Inclined Pipes*. International Journal of Multiphase Flow, Vol. 13, pp. 565–575.

Andritsos, N. and Hanratty, T.J. (1986): *Interfacial Instabilities for Horizontal Gas-Liquid Flows in Pipelines*. International Journal of Multiphase Flow, Vol. 13(5), pp. 583–603.

Andritsos, N. and Hanratty, T.J. (1987): *Influence of Interfacial Waves in Stratified Gas-Liquid Flows*. AIChE Journal, Vol. 33, pp. 444–454.

Arnold, C.R. and Hewitt, G.F. (1967): *Further Developments in the Photography of Two-Phase Gas-Liquid Flow*. Journal of Photographic Science, Vol. 15, pp. 97–114.

Badie, S., Hale, C.P., Lawrence, C.J. and Hewitt, G.F. (2000): *Pressure Gradient and Holdup in Horizontal Two-Phase Gas-Liquid Flows with Low Liquid Loading*. International Journal of Multiphase Flow, Vol. 26, pp. 1525–1543.

Battara, V., Mariani, O., Gentilini, M. and Giachetta, G. (1985): *Condensate Line Correlations for Calculating Holdup, Friction Compared to Field Data*. Oil and Gas Journal, pp. 148–152.

Benkirane, R., Line, A. and Masbernat, L. (1990): *Modelling of Wavy Stratified Flow in a Rectangular Channel. In Phase-Interface Phenomena in Multiphase Flow*. eds G. F. Hewitt, F. Mayinger and J. R. Riznic, Hemisphere, New York., pp. 121–128.

Berthelsen, P.A. (2004a): *A Decomposed Immersed Interface Method for Variable Coefficient Elliptic Equations with Non smooth and Discontinuous Solutions*. Journal of Computational Physics, Vol. 197, pp. 346–386.

Berthelsen, P.A. and Ytrehus, T. (2004): *Numerical Modelling of Stratified Turbulent Two and Three-Phase Pipe Flow with Arbitrary Shaped Interfaces*. The 5th International Conference on Multiphase Flow, Yokohama, Japan, May 30-June 4.

Berthelsen, P.A. and Ytrehus, T. (2005): *Calculation of Stratified Wavy Two-Phase Flow in Pipes*. International Journal of Multiphase Flow, Vol. 31, pp. 571–592.

Biage, M., Delhaye, J.M. and Nakach, R. (1989): *The Flooding Transition: An Experimental Appraisal of the Chaotic Aspect of Liquid Film Flow before the Flooding Point.* Proceedings from Heat Transfer - Philadelphia, pp. 274–279.

Biberg, D. (1999): *Two-phase Stratified Pipe Flow Modelling - A New Expression for the Interfacial Shear Stress.* The 2nd International Symposium on Two-Phase Flow Modelling and Experimentation, Pisa, Italy, May 22-26.

Brackbill, J.U., Kothe, D.B. and Zemach, C. (1992): *A Continuum Method for Modeling Surfaces Tension.* Journal of Computational Physics, Vol. 100, pp. 335–354.

Bradshaw, P., Cebeci, T. and Whitelaw, J. H. (1981): *Engineering Calculation Methods for Turbulent Flow.* Academic Press, London.

Bruno, K. and McCready, M.J. (1989): *Study of the Processes which Control the Interfacial Wave Spectrum in Separated Gas-Liquid Flows.* International Journal of Multiphase Flow, Vol. 15, pp. 531–552.

Busam, S. (2004): *Druckverlust und Wärmeübergang im Entlüftungssystem von Triebwerkslagerkammern.* Dissertation, Institut für Thermische Strömungsmaschinen, Karlsruher Institut für Technologie (KIT).

Busam, S., Glahn, A. and Wittig, S. (2000): *Internal Bearing Chamber Wall Heat Transfer as a function of Operating Conditions and Chamber Geometry.* ASME - Journal of Engineering for Gas Turbines and Power, Vol. 116, pp. 395–401.

Celik, I. and Rodi, W. (1984): *Simulation of Free Surface Effects on Turbulent Channel Flows.* Physico-Chem. J., Vol. 5, pp. 217–227.

Centinbudaklar, A.G. and Jameson, G.J. (1969): *The Mechanism of Flooding in Vertical Countercurrent Two-Phase Flow.* Chem. Eng. Sci., Vol. 24, pp. 1669–1680.

Chandra, B.W. (2006): *Flows in Turbine Engine Oil Sumps.* Dissertation, Purdue University West Lafayette, Indiana, USA.

Chandra, B., Pickering, S., Tittel, M. and Simmons, K. (2010): *Factors Affecting Oil Removal from an Aero engine Bearing Chamber.* ASME-Paper GT2010-22631.

Charnock, H. (1955): *Wind Stress on a Water Surface.* Q. J. Roy. Meteorol. Soc. 81, pp. 639–640.

Chew, J.W. (1996): *Analysis of the Oil Film on a Bearing Chamber Housing.* ASME-96-GT-300.

Cohen, L.S. and Hanratty, T.J. (1965): *Generation of Waves in the Concurrent Flow of Air and a Liquid.* AIChE Journal, Vol. 11, pp. 138–144.

Colebrook, C.F. (1939): *Turbulent Flow in Pipes, with Particular Reference to the Transition Region between the Smooth and Rough Pipe laws.* Journal of the Institution of Civil Engineers, Vol. 11, pp. 133–156.

Cornish, V. (1910): *Waves of the Sea and Other Water Waves*. T. Fisher Unwin, London, England.

Corradini, M.L. (1997): *Fundamentals of Multiphase Flow*. Department of Engineering Physics,University of Wiscosin. Madison WI 53706.

Davidson, L. (2003): *An Introduction to Turbulence Models. Department of Thermo and Fluid Dynamic.* Chalmers University of Technology, Göteborg, Sweden.

De Sampaio, P.A.B., Fuccini, J.L.H. and Su, Jian. (2008): *Modelling of Stratified Gas-liquid Two-Phase Flow in Horizontal Circular Pipes.* International Journal of Heat and Mass Transfer, Vol. 51, pp. 2752–2762.

Dickenson, P. (2009): *The feasibility of smoothed particle hydrodynamics for multiphase oilfield systems.* Seventh International Conference on CFD in the Minerals and Process Industries, CSIRO, Melbourne, Australia, December 9-11.

Drew, D. and Passman, S.L. (1999): *Theory of Multicomponent Fluids.* Springer, New York, Heidelberg.

Dykhno, L.A., Williams, L.R. and Hanratty, T.J. (1994): *Maps of Mean Gas Velocity for Stratified Flows with and without Atomization.* International Journal of Multiphase Flow, Vol. 20, pp. 691–702.

Ebner, J., Schober, P., Schäfer, O., Koch, R. and Wittig., S. (2004): *Modelling of Shear-Driven Liquid Wall Films: Effect of Accelerated Air Flow on the Film Flow Propagation.* Progress in Computational Fluid Dynamics, an International Journal, Vol. 4(3-5), pp. 183–190.

Egorov, Y. (2004): *Contact Condensation in Stratified Steam-Water Flow.* EVOLECORA-D07.

Elsäßer, A. (1998): *Kraftstoffaufbreitung in Verbrennungskraftmaschinen: Grundlagen der Strömung schubspannungsgetriebener Wandfilme.* Dissertation, Institut für Thermische Strömungsmaschinen, Karlsruher Institut für Technologie (KIT).

Espedal, M. (1998): *An Experimental Investigation of Stratified Two-Phase Pipe Flow at Small Inclinations.* Dissertation, Norwegian University of Science and Technology, Department of Applied Mechanics, Thermo- and Fluid Dynamics, Trondheim, Norway.

Fabre, J., Masbernat, L. and Suzanne, C. (1983): *New Results on the Structure of Stratified Gas/Liquid Flow. Advances in Two-phase Flow and Heat Transfer, eds S. Kakac and M. Ishii.* Martinus Nijhoff, The Hague., Vol. 1, pp. 135–150.

Fabre, J., Masbernat, L. and Suzanne, C. (1987): *Stratified flow, Part I: Local Structure.* Multiphase Science and Technology, Hemisphere, Washington D. C., Vol. 3, pp. 285–301.

Farrall, M. (2001): *An Initial Investigation into the Heat Transfer in Shear Driven Thin Films.* Research Report UTC/2001/TF/22/MF, Rolls-Royce UTC, University of Nottingham, UK.

Farrall, M., Hibberd, S. and Simmons, K. (2000): *Computational Modelling of Two-Phase Air/Oil Flow within an Aero-Engine Bearing Chamber*. Proceedings of the ASME FEDSM, Boston, Massachusetts, USA.

Farrall, M., Hibberd, S. and Simmons, K. (2003): *Modelling of Droplet/Film Interaction in an Aero-Engine Bearing Chamber*. ICLASS2003.

Farrall, M., Hibberd, S., Simmons, K., Busam, S., Gorse, P. and Dullenkopf, K. (2004): *A Numerical Model for Oil Film Flow in an Aero-Engine Bearing Chamber and Comparison with Experimental Data*. ASME-Paper GT2004-53698.

Feind, K. (1960): *Strömungsuntersuchungen bei Gegenströmung von Rieselfilmen und Gas in lotrechten Rohren*. VDI Forschungsh, Vol. 26, pp. 481.

Fernandino, M. and Ytrehus, T. (2006): *Determination of Flow Sub-Regimes in Stratified Air-Water Channel Flow Using LDV Spectra*. International Journal of Multiphase Flow, Vol. 32, pp. 436–446.

Ferreira, V.C.F. (2004): *Estudos Hidrodinâmicos e de Sujamento em Condutas Horizontais de Secção Recta Quadrada*. MSc. Thesis, Faculty of Engineering, University of Oporto.

Flores, A.G., Crowe, K.E. and Griffith, P. (1995): *Gas-phase Secondary Flow in Horizontal, Stratified and Annular Two-Phase Flow*. International Journal of Multiphase Flow, Vol. 21, pp. 207–221.

Flouros, M. (2005): *The Impact of Oil and Sealing Air Flow, Chamber Pressure, Rotor Speed and Axial Load on the Power Consumption in an Aeroengine Bearing Chamber*. ASME - Journal of Engineering for Gas Turbines and Power, Vol. 127, pp. 182–186.

Flouros, M. (2008): *Analytical and Numerical Simulation of the Two Phase Flow Heat Transfer in the Vent and Scavenge Pipes of the Clean Engine Demonstrator*. ASME-Paper GT2008-50130.

Fluent 6.3 (2006): *User's Guide*.

Frank, T. (2005): *Numerical Simulation of Slug Flow Regime for an Air-Water Two-Phase Flow in Horizontal Pipes*. The 11th International Topical Meeting on Nuclear Reactor Thermal-Hydraulics (NURETH-11).

Friedel, L. (1977): *Momentum Exchange and Pressure Drop in Two-Phase Flow*. Proceedings of NATO Advanced Study Institute (Edited by S. Kakaç & F. Mayinger), pp. 239–312.

Gargallo, M., Schulenberg, T., Mayer, L. and Laurien, L. (2005): *Counter-Current Flow Limitations During Hot Leg Injection in Pressurized Water Reactors*. Nuclear Engineering and Design, Vol. 235, pp. 785–804.

Gerendás, M., Samenfink, W. and Wittig, S. (1995): *Experimental and Numerical Investigations on Evaporating Multi-Component Liquid Films in Turbulent Air Flow*. Flows with Phase Transition EUROMECH Colloquium 331, DLR-Mitteilungen 94-11.

Ghiaasiaan, S.M., Wu, X., Sadowski, D.L. and Abdel-Khalik, S.I. (1997): *Hydrodynamic Characteristics of Counter-Current Two-Phase Flow in Vertical and Inclined Channels: Effects of Liquid Properties*. Internation Journal of Multiphase Flow, Vol. 23(6), pp. 1063–1083.

Ghiaasiaan, S. M., Taylor, K. E., Kamboj, B. K. and Abdel-Khalik, S. I. (1995a): *Countercurrent Two-Phase Flow Regimes and Void Fraction in Vertical and Inclined Channels*. Nuclear Science Engineering, Vol. 119, pp. 182–194.

Ghorai, S. and Nigam, K.D.P. (2005): *CFD Modeling of Flow Profiles and Interfacial Phenomena in Two-Phase Flow in Pipes*. Chemical Engineering Process, Vol. 45, pp. 55–65.

Glahn, A. and Wittig, S. (1996): *Two-Phase Air/Oil Flow in Aero Engine Bearing Chambers: Characterization of Oil Film Flow*. ASME Journal of Engineering for Gas Turbines and Power, Vol. 118(3), pp. 578–583.

Glahn, J.A. (1995): *Zweiphasenströmungen in Triebwerkslagerkammern - Charakterisierung der Ölfilmströmung und des Wärmeübergangs*. Dissertation, Institut für Thermische Strömungsmaschinen, Karlsruher Institut für Technologie (KIT).

Gorse, P. (2007): *Tropfenentstehung und Impulsaustausch in Lagerkammern von Flugtriebwerken*. Dissertation, Institut für Thermische Strömungsmaschinen, Karlsruher Institut für Technologie (KIT).

Gorse, P., Willenborg, K., Busam, S., Ebner, J., Dullenkopf, K. and Wittig, S. (2003): *3D-LDA Measurements in an Aero-Engine Bearing Chamber*. ASME-Paper GT2003-38376.

Gorse, P., Busam, S. and Dullenkopf, K. (2004): *Influence of Operating Condition and Geometry on the Oil Film Thickness in Aero-Engine Bearing Chambers*. ASME Journal of Engineering for Gas Turbines and Power, Vol. 128(1), pp. 103–110.

Gorse, P., Dullenkopf, K., Bauer, H.-J. and Wittig, S. (2008): *An Experimental Study on Droplet Generation in Bearing Chambers Caused by Roller Bearings*. ASME-Paper GT2008-51281.

Govan, A.H., Hewitt, G.F., Richter, H.J. and Scott, A. (1991): *Flooding and Churn Flow in Vertical Pipes*. International Journal of Multiphase Flow, Vol. 17, pp. 27–44.

Gregory, G. A. and Fogarasi, M. (1985): *A Critical Evaluation of Multiphase gas-Liquid Pipeline Calculation Methods*. Proceedings of the Second International Conference on Multiphase Flow, London, UK, pp. 93–108.

Hanratty, T.J. and Engen, J.M. (1957): *Interaction between a Turbulent Air Stream and a Moving Water Surface*. AIChE Journal, Vol. 3, pp. 299–304.

Hanratty, T.J. and Hershman, A. (1961): *Initiation of Roll Waves*. AIChE Journal, Vol. 7(3), pp. 488–497.

Hart, J., Hamersma, P.J. and Fortuin, J.M.H. (1989): *Correlations Predicting Frictional Pressure Drop and Liquid Holdup during Horizontal Gas-Liquid Pipe Flow with a Small Liquid Holdup*. International Journal of Multiphase Flow, Vol. 15(6), pp. 947–964.

Hewitt, G.F. and Wallis, G.B. (1963): *Flooding and Associated Phenomena in Falling Film Flow in a Vertical Tube.* UKAEA Report, AERE R-4022.

Höfler, C., Braun, S., Koch, R. and Bauer, H.-J. (2011): *Towards the Numerical Prediction of Primary Atomization Using Smoothed Particle Hydrodynamics.* 24th European Conference on Liquid Atomization and Spray Systems, Estoril, Portugal, September 5-7.

Himmelsbach, J. (1992): *Zweiphasenströmungen mit schubspannungsgetriebenen welligen Flüssigkeitsfilmen in turbulenter Heißluftströmung - Meßtechnische Erfassung und numerische Beschreibung.* Dissertation, Institut für Thermische Strömungsmaschinen, Karlsruher Institut für Technologie (KIT).

Himmelsbach, J., Noll, B. and Wittig, S. (1994): *Experimental and Numerical Studies of Evaporating Wavy Fuel Films in Turbulent Air Flow.* International Journal of Heat and Mass Transfer, Vol. 37, pp. 1217–1226.

Hirt, C.W. and Nichols, B.D. (1981): *Volume of Fluid (VOF) Method for the Dynamics of Free Boundaries.* Journal of Computational Physics, Vol. 39, pp. 201–225.

Imura, H., Kusuda, H. and Funatsu, S. (1977): *Flooding Velocity in a Countercurrent Annular Two-Phase Flow.* Chemical Engineering Science, Vol. 32, pp. 79–87.

Ishii, M. (1975): *Thermo-Fluid Dynamic Theory of Two-Phase Flow.* Eyrolles, Paris.

Issa, R.I. (1988): *Prediction of Turbulent, Stratified, Two-Phase Flow in Inclined Pipes and Channels.* International Journal of Multiphase Flow, Vol. 14, pp. 141–154.

Jeffreys, H. (1925): *On the Formation of Water Waves by Wind.* Proceedings of the Royal Society of England, Series A, Vol. 107(742), pp. 189–201.

Johns, D. M., Theofanous, T.G. and Houze, R.N. (1975): *Turbulent Characteristics of Two-Phase, Gas-Liquid Stratified Channel Flow.* Proceedings of the Third Symposium on Turbulence in Liquids, eds G. K. Patterson and J. L. Zakin, pp. 250–258.

Kihm, K.D. (1996): *Investigation of the Effect of Coal Particle Sizes on the Interfacial and Rheological Properties of Coal-Water Slurry Fuels.* Quarterly report no. 2 (FY96), Department of Mechanical Engineer, Texas A&M University.

Klingsporn, M. (2004): *Advanced Transmissions and Oil System Concepts for Modern Aero-Engines.* ASME-Paper GT2004-53578.

Launder, B.E. and Spalding, D.B. (1974): *The Numerical Computation of Turbulent Flows.* Computational Methods in Applied Mechanics and Engineering, Vol. 3, pp. 269–289.

Lee, S.C. (1992): *Interfacial Friction Factors in Countercurrent Stratified Two-Phase Flow.* Chemical Engineering Communications, Vol. 118, pp. 3–16.

Lee, S.C. and Bankoff, S.G. (1982): *Stability of Steam-Water Countercurrent Flow in an Inclined Channel.* ASME-Paper 82-WA/HT-6.

Lin, P.Y. and Hanratty, T.J. (1987): *Effect of Pipe Diameter on Flow Patterns for Air-Water Flow in Horizontal Pipes*. International Journal of Multiphase Flow, Vol. 13, pp. 549–563.

Liovic, P. and Lakehal, D. (2007): *Multi-physics Treatment in the Vicinity of Arbitrarily Deformable Gas-Liquid Interfaces*. Journal of Computational Physics, Vol. 222(2), pp. 504–535.

Lipp, S. (2003): *Charakterisierung der Zweiphasigen Öl/Luftströmung in einer Lagerkammer unter triebwerksnahen Bedingungen*. Studienarbeit, Institut für Thermische Strömungsmaschinen, Karlsruher Institut für Technologie (KIT).

Lockhart, R. W. and Martinelli, H. C. (1949): *Proposed Correlation of Data for Isothermal Two Phase, Two Component Flow in Pipes*. Chemical Engineering Progress, Vol. 45, pp. 39–48.

Lopez, D. (1994): *Ecoulements diphasiques a phases separees a faible contenu de liquide*. Dissertation, Institut National Polytechnique de Toulouse, Toulouse, France.

Lorencez, C., Chang, T. and Kawaji, M. (1991): *Investigation of Turbulent Momentum Transfer at a Gas-Liquid Interface in a Horizontal Countercurrent Stratified Flow*. Proceedings, 1st ASME/JSME Fluids Engineering Conference, ASME-FED, Vol. 110, pp. 97–102.

Lorencez, C., Kawaji, M., Ojha, M., Ousaka, A. and Murao, Y. (1993): *Application of a Photochromic Dye Activation Method to Stratified Flow with Smooth and Wavy Gas-Liquid Interface*. ANS Proceedings of National Heat Transfer Conference, Vol. 7, pp. 160–168.

Lorencez, C., Nasr-Esfahany, M., Kawaji, M. and Ojha, M. (1997): *Liquid Turbulence Structure at a Sheared and Wavy Gas-Liquid Interface*. International Journal of Multiphase Flow, Vol. 23, pp. 205–226.

Mandhane, J.M., Gregory, C.A. and Aziz, K. (1974): *A Flow Pattern Map for Gas-Liquid Flow in Horizontal Pipes*. Chemical Engineering Progress, Vol. 45, pp. 39–40.

Maqableh, A., Simmons, K., Hibberd, S., Power, H. and Young, C. (2003): *CFD Modelling of Three-Component Air/Oil Flow and Heat Transfer in a Rotating Annulus*. 11th International-Conference on CFD, CFD Society of Canada.

McKee, S., Tome´, M.F., Ferreira, V.G., Cuminato, J.A., Castelo, A., Sousa, F.S. and Mangiavacchi, N. (2008): *Review - The MAC method*. Computers & Fluids, Vol. 37, pp. 907–930.

Meknassi, F., Benkirane, R., Line´, A. and Masbernat, L. (2000): *Numerical Modeling of Wavy Stratified Two-Phase Flow in Pipes*. Chemical Engineering Science, Vol. 55, pp. 4682–4697.

Monaghan, J.J. (2011): *A Turbulence Model for Smoothed Particle Hydrodynamics*. European Journal of Mechanics B/Fluids, Vol. 30, pp. 360–370.

Morrision, G.L., Johnson, M.C. and Tatterson, G.B. (1991): *Three Dimensional Laser Anemometer Measurements in a Labyrinth Seal*. ASME Journal of Engineering for Gas Turbines and Power, Vol. 113, pp. 119–125.

Mouza, A.A., Paras, S.V. and Karabelas, A.J. (2001): *CFD Code Application to Wavy Stratified Gas-Liquid Flow*. Trans IChemE, Vol. 79(Part A), pp. 561–658.

Murata, A., Hihara, E. and Saito, T. (1991): *Turbulence below an Air-Water Interface in a Horizontal Channel*. Proceedings, 1st ASME/JSME Fluids Engineering Conference, ASME-FED, Vol. 110, pp. 67–74.

Newton, C.H. and Behnia, M. (2000): *Numerical Calculation of Turbulent Stratified Gas-Liquid Pipe Flows*. International Journal of Multiphase Flow, Vol. 26, pp. 327–337.

Newton, C.H. and Behnia, M. (2001): *A Numerical Model of Stratified Wavy Gas-Liquid Pipe Flow*. Chemical Engineering Science, Vol. 56, pp. 6851–6861.

Omgba-Essama, C. (2004): *Numerical Modelling of Transient Gas-Liquid Flows (Application to Stratified & Slug Flow Regimes)*. PhD Thesis, School of Engineering, Applied Mathematics and Computing Group, Cranfield University, UK.

Ottens, M., Hoefsloot, H.C.J. and Hamersma, P.J. (2001): *Correlations Predicting Liquid Hold-up and Pressure Gradient in Steady-State (nearly) Horizontal Co-Current Gas-Liquid Pipe Flow*. Trans IChemE, Vol. 79(5), pp. 581–592.

Paras, S.V. and Karabelas, A.J. (1992): *Measurements of Local Velocities inside Thin Liquid Films in Horizontal Two-Phase Flow*. Experiments in Fluids, Vol. 13, pp. 190–198.

Paras, S.V., Vlachos, N.A. and Karabelas, A.J. (1994): *Liquid Layer Characteristics in Stratified Atomization Flow*. International Journal of Multiphase Flow, Vol. 20, pp. 939–956.

Paras, S.V., Vlachos, N.A. and Karabelas, A.J. (1998): *LDA Measurements of Local Velocities Inside the Gas Phase in Horizontal Stratified Atomization Two-Phase Flow*. International Journal of Multiphase Flow, Vol. 24, pp. 651–661.

Patel, V.C., Rodi, W. and Scheuerer, G. (1985): *Turbulence Models for Near-Wall and Low Reynolds Number Flows - A Review*. AIAA Journal, Vol. 23(9), pp. 1308–1319.

Pawloski, J.L., Ching, C.Y. and Shoukri, M. (2004): *Measurement of Void Fraction and Pressure Drop of Air-Oil Two-Phase Flow in Horizontal Pipes*. ASME Journal of Engineering for Gas Turbines and Power, Vol. 126, pp. 107–118.

Peduto, D., Kurz, W. and Dullenkopf, K. (2010): *Film Thickness Measurements in Bearing Chamber Test Rig*. ITS Internal Report.

Petalas, N. and Aziz, K. (1998): *A Mechanistic Model for Multiphase Flow in Pipes*. 49th Annual Technical Meeting of the Petroleum Society of the Canadian Institute of Mining, Metallurgy and Petroleum.

Rashidi, M. and Banerjee, S. (1990): *Streak Characteristics and Behavior Near Wall and Interface in Open Channel Flows*. ASME Journal of Fluids Engineering, Vol. 112, pp. 164–170.

Reboux, S., Sagaut, P. and Lakehal, D. (2006): *Large-Eddy Simulation of Sheared Interfacial Flow*. Physics of Fluids, Vol. 18(10), pp. 105105–105120.

Revellin, R. (2005): *Experimental Two-Phase Fluid Flow in Microchannels*. Dissertation, Ecole polytechnique Fédérale de Lausanne, Switzerland.

Ribeiro, A.M., Ferreira, V. and Campos, J.B.L.M. (2006): *On the Comparison of New Pressure Drop and Hold-up Data for Horizontal Air-Water Flow in a Square Cross-section Channel Against Existing Correlations and Models*. International Journal of Multiphase Flow, Vol. 32, pp. 1029–1036.

Rider, W.J. and Kothe, D. (1998): *Reconstructing volume tracking*. Journal of Computational Physics, Vol. 141, pp. 112–152.

Robinson, A., Eastwick, C. and Morvan, H. (2008): *Computational Investigations into Aero-engine Bearing Chamber Off-take Flows*. ASME-Paper GT2008-50634.

Robinson, A., Eastwick, C. and Morvan, H. (2010): *Further Computational Investigations into Aero-engine Bearing Chamber Off-take Flows*. ASME-Paper GT2010-22626.

Rosant, J.M. (1983): *Ecoulements diphasiques liquide-gaz en conduite circulaire, étude de la configuration stratifiée au voisinage de l'horizontale*. Dissertation, École nationale supérieure de mécanique de Nante, France.

Rosskamp, H., Willmann, M. and Wittig, S. (1997a): *Heat up and Evaporation of Shear-driven Liquid Wall Films in Hot Turbulent Air Flow*. Proceedings of the 2nd International Symposium on Turbulence, Heat and Mass Transfer, Delft, Nederland, June 9-12.

Rosskamp, H., Willmann, M. and Wittig, S. (1997b): *Computation of Two-Phase Flows in Low-NOX Combustor Premix Ducts Utilizing Fuel Film Evaporation*. ASME-Paper 97-GT-226.

Rosskamp, H., Elsäßer, A., Samenfink, W., Meisl, J., Willmann, M. and Wittig, S. (1998): *An Enhanced Model for Predicting the Heat Transfer to Wavy Shear-Driven Liquid Wall Films*. Third International Conference on Multiphase Flow, ICMF Lyon, France, June 8-12.

Roy, R.P. and Jain, S. (1989): *A Study of Thin Water Film Flow Down an Inclined Plate without and with Countercurrent Air Flow*. Experiments in Fluids, Vol. 7, pp. 318–328.

Russel, T.W.F., Etchells, A.W., Jensen, R.H. and Arruda, P.J. (1974): *Pressure Drop and Holdup in Stratified Gas-Liquid Flow*. AIChE Journal, Vol. 30, pp. 664–669.

Samenfink, W., Elsäßer, A., Wittig, S. and Dullenkopf, K. (1996): *Internal Transport Mechanism of Shear-Driven Liquid Films*. International Symposium on Applications of Laser Techniques to Fluid Mechanics, Lisbon, Portugal, July 8-11.

Sattelmayer, T. and Wittig, S. (1986): *Internal Flow Effects in Prefilming Airblast Atomizers: Mechanisms of Atomization and Droplet Spectra*. ASME Paper 86-GT-150.

Schmidt, J., Hank, W.K., Klein, A. and Maier, K. (1981): *The Oil/Air System of a Modern Fighter Aircraft Engine*. AGARD-CP-323.

Schmälzle, C. (1997): *Charakterisierung der Ölfilmströmung in Triebwerkslagerkammern - Strömungsvisualisierung und Filmgeschwindigkeitsmessungen mit Hilfe der Laser Doppler Anemometer*. Studienarbeit, Institut für Thermische Strömungsmaschinen, Karlsruher Institut für Technologie (KIT).

Schober, P. (1998): *Bewertung unterschiedlicher Gestalungskonzepte am Eintritt von Abluft und Absaugleitungen in Lagerkammern*. Diplomarbeit, Institut für Thermische Strömungsmaschinen, Karlsruher Institut für Technologie (KIT).

Schober, P., Ebner, J., Schäfer, O. and Wittig, S. (2003): *Experimental Study on the Effect of a Strong Negative Pressure Gradient on a Shear-Driven Liquid Fuel Film*. 9th International Conference on Liquid Atomization and Spray Systems (ICLASS), Paper 8-18, Sorrento, Italy, July 13-18.

Schramm, V., Denecke, J., Kim, S. and Wittig, S. (2004): *Shape Optimization of a Labyrinth Seal Applying the Simulated Annealing Method*. International Journal of Rotating Machinery, Vol. 10(5), pp. 365–371.

Shi, J. and Kocamustafaorgullari, G. (1994): *Interfacial Measurements in Horizontal Stratified Flow Patterns*. Nuclear Engineering and Design, Vol. 149, pp. 81–96.

Shim, W.J. and Jo, C.H. (2000): *Analysis of Pressure Fluctuations in Two-phase Vertical Flow in Annulus*. Journal of Industrial and Engineering Chemistry, Vol. 6(3), pp. 167–173.

Shimo, M., Canino, J.V. and Heister, S.D. (2005): *Modeling Oil Flows on Seal Runners and Engine Sump Walls*. ASME Journal of Engineering for Gas Turbines and Power, Vol. 127(4), pp. 827–835.

Shoham, O. and Taitel, Y. (1984): *Stratified Turbulent-Turbulent Gas-Liquid Flow in Horizontal and Inclined Pipes*. AIChE Journal, Vol. 30, pp. 377–385.

Sill, K. H. (1980): *Experimentelle Bestimmung der Grenzflächenstruktur und der mittleren Filmdicke von strömenden Flüssigkeitsfilmen mit Hilfe einer Lichtabsorptionsmethode*. Forschung in der Kraftwerkstechnik, Sammelband der VGB-Tagung, Essen 1980, pp. 232–338.

Stäbler, T., Meyer, L., Schulenberg, T. and Laurien, E. (2005): *Investigations on Flow Reversal in Stratified Horizontal Flow*. Proceedings of International Conference on Nuclear Energy for New Europe, Bled, Slovenia.

Stäbler, T.D., Meyer, L., Schulenberg, T. and Laurien, E. (2006): *Turbulence and Void Distribution in Horizontal Counter-Current Stratified Flow*. Proceedings of 17th International Symposium on Transport Phenomena, Toyama, Japan.

Steen, D.A. and Wallis, G.B. (1964): *The Transition from Annular to Annular-Mist Cocurrent Two-Phase Down Flow*. AEC Report NYO-3114-2.

Strakey, P. (2004): *Application of Multiphase Flow in Rocket Injector science*. National Energy Technology Laboratory, Morgantown, WV and Douglas Talley, U.S. Air Force Research Laboratory, Edwards Air Force Base, CA.

Strand, O. (1993): *An Experimental Investigation of Stratified Two-Phase Flow in Horizontal Pipes*. Dissertation, University of Oslo, Norway.

Taitel, Y. and Dukler, A.E. (1976): *A Model for Predicting Flow Regime Transitions in Horizontal and Near Horizontal Gas-Liquid Flow*. AIChE Journal, Vol. 22, pp. 47–55.

Tennekes, H. and Lumley, J. L. (1972): *A First Course in Turbulence*. MIT Press, Cambridge, MA.

Thome, J. R. (2004): *Wolverine Engineering Data Book III*. Wolverine Tube, Inc.

Tribbe, C. and Muller-Steinhagen, H.M. (2000): *An Evaluation of the Performance of Phenomenological Models for Predicting Pressure Gradient during Gas-Liquid Flow in Horizontal Pipelines*. International Journal of Multiphase Flow, Vol. 26, pp. 1019–1036.

Vallée, C., Höhne, T., Prasser, H-M. and Sühnel, T. (2008): *Experimental investigation and CFD simulation of horizontal stratified two-phase phenomena*. Nuclear Engineering and Design, Vol. 238, pp. 637–646.

Versteeg, H.K. and Malalasekera, W. (1995): *An Introduction to Computational Fluid Dynamics - The Finite Volume Method*. First edition, Longman Group Ltd., Essex, UK.

Vijayan, M., Jayanti, S. and Balakrishnan, A.R. (2001): *Experimental Study of Air-Water Countercurrent Annular Flow under Post-Flooding Conditions*. International Journal of Multiphase Flow, Vol. 28, pp. 51–67.

Vlachos, N.A., Paras, S.V. and Karabelas, A.J. (1997): *Liquid-to-Wall Shear Stress Distribution in Stratified Atomisation Flow*. International Journal of Multiphase Flow, Vol. 23, pp. 845–863.

Vorobyev, A., Kriventsev, V. and Maschek, W. (2010): *Analysis of Central Sloshing Experiment using Smoothed Particle Hydrodynamics (SPH) Method*. 18th International Conference on Nuclear Engineering, ICONE18-29805, Vol. 4, pp. 751–759.

Wang, Y., Simmons, K. and Hibberd, S. (2000): *Consideration of Oil Film Motion on the Surface of Rotating Shafts within the HP-IP Bearing Chamber*. Research Report UTC/1998/TF/19/YW, Rolls-Royce UTC, University of Nottingham, UK.

Weissman, J., Duncan, D., Gibson, J. and Crawford, T. (1979): *Effects of Fluid Properties and Pipe Diameter on Two-Phase Patterns in Horizontal Lines*. International Journal of Multiphase Flow, Vol. 5, pp. 437–462.

Wintterle, T. (2008): *Modellentwicklung und numerische Analyse zweiphasig geschichteter horizontaler Strömungen*. Dissertaion, Faculty of Mechanical Engineering, University Stuttgart, Germany.

Wintterle, T., Laurien, E., Stäbler, T., Meyer, L. and Schulenberg, T. (2007): *Experimental and Numerical Investigation of Counter-Current Stratified Flows in Horizontal Channels.* Nuclear Engineering and Design, Vol. 238, pp. 627–636.

Wittig, S., Himmelsbach, J., Noll, B., Feld, H.J. and Samenfink, W. (1992): *Motion and Evaporation of Shear-Driven Liquid Films in Turbulent Gases.* ASME - Journal of Engineering for Gas Turbines and Power, Vol. 114, pp. 395–400.

Wittig, S., Glahn, A. and Himmelsbach, J. (1994): *Influence of High Rotational Speeds on Heat Transfer and Oil Film Thickness in Aero Engine Bearing Chambers.* ASME - Journal of Engineering for Gas Turbines and Power, Vol. 116, pp. 395–401.

Wurz, D. E. (1976): *Experimental Investigation into the Flow Behaviour of Thin Water Films; Effect on a Cocurrent Air Flow of Moderate to High Supersonic Velocities: Pressure Distribution at the Surface of Rigid Wavy Reference Structures.* Archiwum Mechaniki Stosowanej, Warsaw, Poland, Vol. 28(5-6), pp. 969–987.

Yamaguchi, K. and Yamazaki, K. (1984): *Combinated Flow Pattern Map for Cocurrent and Countercurrent Air-Water Flows in Vertical Tubes.* Journal of Nuclear Science and Technology, Vol. 21, pp. 321–327.

Yamaguchi, K. and Yamazaki, Y. (1982): *Characteristics of Countercurrent Gas-Liquid Two-Phase Flow in Vertical Tubes.* Journal of Nuclear Science and Technology, Vol. 19, pp. 985–996.

Young, C. and Chew, J. W. (2005): *Evaluation of the Volume of Fluid Modelling Approach for Simulation of Oil/Air System Flows.* ASME-Paper GT2005-68861.

Zabaras, G.J. and Dukler, A.E. (1988): *Countercurrent Gas-Liquid Annular Flow Including the Flooding State.* AIChE Journal, Vol. 34, pp. 389–396.

Zimmermann, H., Kammerer, A., Fischer, R. and Rebhan, D. (1991): *Two-Phase Correlations in Air/Oil Systems of Aero-Engines.* International Gas Turbine and Aero-Engine Congress and Exposition, Florida, USA.

Lebenslauf

Name	Amir Aleem Hashmi
Geburtsdatum	24. Juli 1981
Geburtsort	Karachi, Pakistan
Familienstand	verheiratet

Schulausbildung:

1986 – 1997	PAF Degree College, Shara-e-Faisal, Karachi, Pakistan
1997 – 1999	Government Science College (Pre-Engineering), Malir Cantt, Karachi, Pakistan

Studium:

01/2000 – 06/2003	Bachelorstudiengang des allgemeinen Maschinenbaus an der National University of Sciences & Technolgy (NUST), PNEC, Karachi, Pakistan
10/2003 – 05/2006	Masterstudiengang Simulation Techniques in Mechanical Engineering an der RWTH Aachen

Berufliche Tätigkeit:

Seit 04/2007	Wissenschaftlicher Angestellter am Institut für Thermische Strömungsmaschinen des Karlsruher Instituts für Technologie (KIT)